Everyday Mathematics®

The University of Chicago School Mathematics Project

Student Math Journal
Volume 2

Grade 2

Chicago, IL • Columbus, OH • New York, NY

The University of Chicago School Mathematics Project (UCSMP)
Max Bell, Director, UCSMP Elementary Materials Component; Director, *Everyday Mathematics* First Edition; James McBride, Director, *Everyday Mathematics* Second Edition; Andy Isaacs, Director, *Everyday Mathematics* Third Edition; Amy Dillard, Associate Director, *Everyday Mathematics* Third Edition; Rachel Malpass McCall, Associate Director, *Everyday Mathematics* Common Core State Standards Edition

Authors
Max Bell, Andy Isaacs, Jean Bell, James McBride, John Bretzlauf, Cheryl G. Moran, Amy Dillard, Kathleen Pitvorec, Robert Hartfield, Peter Saecker

Technical Art
Diana Barrie

Third Edition Teachers in Residence
Kathleen Clark, Patti Satz

UCSMP Editorial
John Wray, Don Reneau

Contributors
Mikhail Guzowski, Robert Balfanz, Judith Busse, Mary Ellen Dairyko, Lynn Evans, James Flanders, Dorothy Freedman, Nancy Guile Goodsell, Pam Guastafeste, Nancy Hanvey, Murray Hozinsky, Deborah Arron Leslie, Sue Lindsley, Mariana Mardrus, Carol Montag, Elizabeth Moore, Kate Morrison, William D. Pattison, Joan Pederson, Brenda Penix, June Ploen, Herb Price, Dannette Riehle, Ellen Ryan, Marie Schilling, Susan Sherrill, Patricia Smith, Robert Strang, Jaronda Strong, Kevin Sweeney, Sally Vongsathorn, Esther Weiss, Francine Williams, Michael Wilson, Izaak Wirzup

Photo Credits
Cover (l)Linda Lewis/Frank Lane Picture Agency/CORBIS, (r)C Squared Studios/Getty Images, (bkgd)Estelle Klawitter/CORBIS; **Back Cover Spine** C Squared Studios/Getty Images; **iii** (t)The McGraw-Hill Companies, (b)Photodisc/Getty Images; **iv** (t)The McGraw-Hill Companies, (b)Photolibrary; **v** (t)Stockbyte/Getty Images, (b)Siede Preis/Getty Images; **vi vii** The McGraw-Hill Companies; **viii** Image Source/Alamy; **others** The McGraw-Hill Companies.

---

everyday**math**.com

STEM McGraw-Hill is committed to providing instructional materials in Science, Technology, Engineering, and Mathematics (STEM) that give all students a solid foundation, one that prepares them for college and careers in the 21st century.

Send all inquiries to:
McGraw-Hill Education
STEM Learning Solutions Center
P.O. Box 812960
Chicago, IL 60681

ISBN: 978-0-07-657640-1
MHID: 0-07-657640-X

Printed in the United States of America.

10 11 12 13 QVS 17 16 15 14

Contents

UNIT 7 Patterns and Rules

UNIT 8 Fractions

UNIT 9 Measurement

UNIT 10 Decimals and Place Value

UNIT 11 Whole Number Operations Revisited

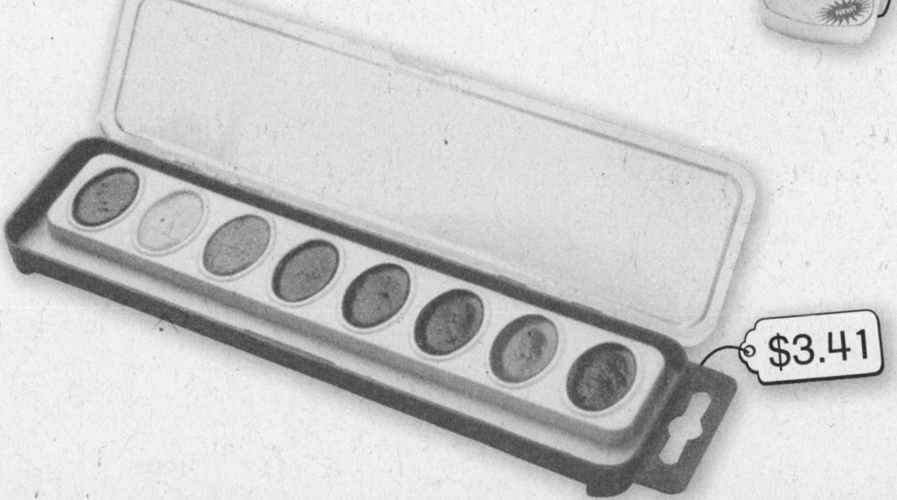

UNIT 12 Year-End Reviews and Extensions

Projects

Activity Sheets

Date ______________________ Time ______________________

LESSON 7•1 Using a Calculator to Find Patterns

1. Use a calculator to count by 5s starting with the number 102. Color the counts on the grid with a crayon. Look for a pattern.

									100
101	102	103	104	105	106	107	108	109	110
111	112	113	114	115	116	117	118	119	120
121	122	123	124	125	126	127	128	129	130

2. Pick a number to count by. Start with a number less than 310. Use your calculator to count. Record your counts on the grid with a crayon.

									300
301	302	303	304	305	306	307	308	309	310
311	312	313	314	315	316	317	318	319	320
321	322	323	324	325	326	327	328	329	330
331	332	333	334	335	336	337	338	339	340
341	342	343	344	345	346	347	348	349	350
351	352	353	354	355	356	357	358	359	360
361	362	363	364	365	366	367	368	369	370

I counted by 5 starting with the number 305.

Here is a pattern that I found: 305 310 315 320 325 330

335 340 345 350 350 355 360 365 370 3[illegible] 390

____

Date Time

LESSON 7·1 Math Boxes

1. Which one is certain to happen? Circle the best answer.

   A. A spaceship will land at school.

   B. Your favorite sports team will win every time.

   C. Spring will follow winter.

   D. You will be a movie star.

2. Solve.

   Unit

   17 − 9 = 8

   27 − 9 = 18

   57 − 9 = 48

   68 = 77 − 9

   88 = 97 − 9

3. Make a 7-by-7 array with dots.

   How many in all? 49 dots

4. Arrange the allowances in order from the minimum (smallest) to the maximum (largest).

   $10, $3, $7, $1, $4

   1, 3, 4, 7, 10

   The minimum is 1.

   The maximum is 10.

5. Match each person with the correct weight.

newborn	about 144 pounds
2nd grader	about 63 pounds
adult	about 7 pounds

6. How many boxes are on this Math Boxes page?

   6 boxes

   How many boxes are on $\frac{1}{2}$ of this page?

   [illegible] boxes

Date Time

LESSON 7•2

Making 10s

Record three rounds of *Hit the Target.*

Example Round:

Target number: 40

Starting Number	Change	Result	Change	Result	Change	Result
12	+38	50	−10	40		

Round 1

Target number: 40

Starting Number	Change	Result	Change	Result	Change	Result
10	+70	80	−10	70		

Round 2

Target number: 70

Starting Number	Change	Result	Change	Result	Change	Result
10	+70	80	−20	60		

Round 3

Target number: 60

Starting Number	Change	Result	Change	Result	Change	Result

Date Time

LESSON 7·2

Solving Subtraction Problems

Use base-10 blocks to help you subtract.

1.

longs	cubes
5	6
− 3	9
2	7

2.

longs	cubes
7	3
− 1	4
6	1

Use any strategy to solve.

3. Ballpark estimate: 28

$$\begin{array}{r} 47 \\ -19 \\ \hline 28 \end{array}$$

4. Ballpark estimate: 65

$$\begin{array}{r} 88 \\ -23 \\ \hline 65 \end{array}$$

5. Ballpark estimate: 17

$$\begin{array}{r} 82 \\ -65 \\ \hline 17 \end{array}$$

6. Ballpark estimate: 122

$$\begin{array}{r} 160 \\ -38 \\ \hline 122 \end{array}$$

Date Time

LESSON 7·2

Math Boxes

1. 24 children. 6 in each row. Draw an array.

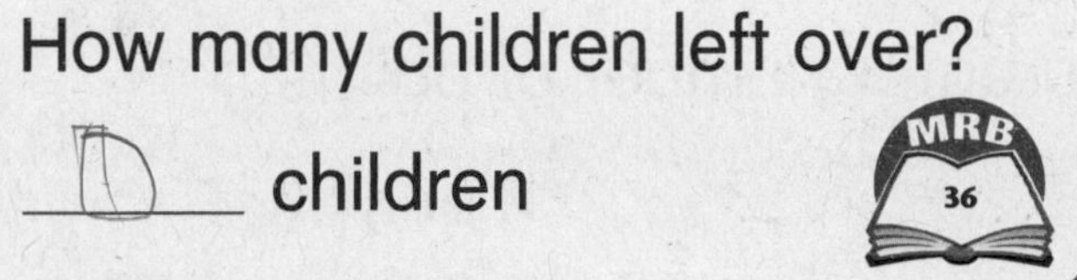

How many rows? ______ rows

How many children left over?

______ children

MRB 36

2. 15 dogs.
13 cats.
12 birds.

How many animals?

______ animals

3. Fill in the missing numbers.

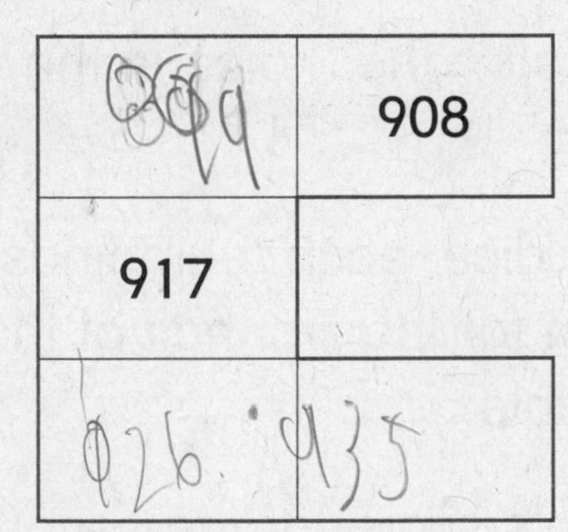

	908
917	

4. Draw a line of symmetry on this triangle.

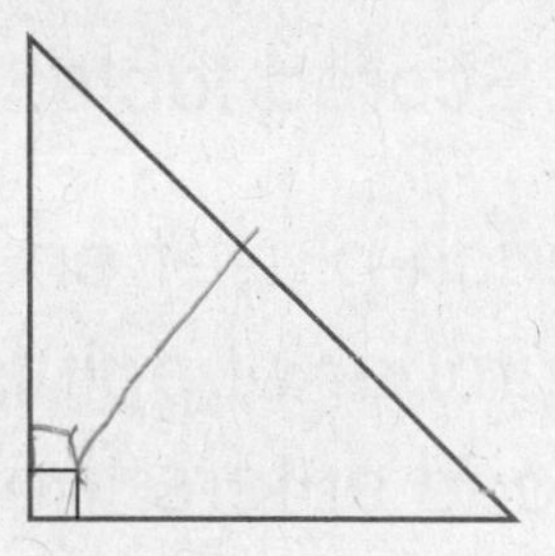

5. Find the rule. Complete the table.

in	out
193	183
232	222
441	
	346

6. ______ of the shape is shaded. Circle the best answer.

A. 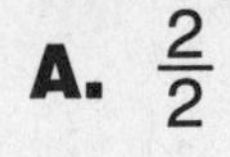 $\frac{2}{2}$

B. 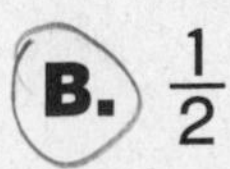 $\frac{1}{2}$

C. 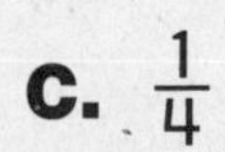 $\frac{1}{4}$

D. 0

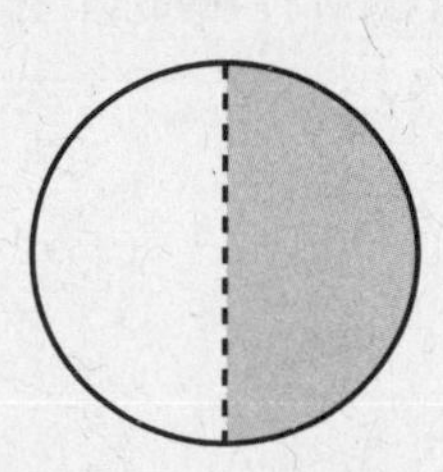

MRB 13

LESSON 7•3

Playing *Basketball Addition*

Materials

- ☐ *Basketball Addition* scoreboard (*Math Journal 2,* p. 167 or *Math Masters,* p. 451)
- ☐ 3 regular dice

Players 2 teams of 3–5 players each

Skill Add three or more 1- and 2-digit numbers

Object of the Game To score a greater number of points

Directions

1. Players on opposite teams take turns rolling the 3 dice.
2. Each player enters the sum of the numbers on the 3 dice in the Points Scored table.
3. After each player on a team has rolled the dice, each team finds the total number of points scored by their team for the first half of the game and enters the Team Score in the table.
4. Players repeat Steps 1–3 to find their team's score for the second half of the game.
5. Each team adds their team totals from both halves of the game to find their team's final score.
6. The team with the greater number of points wins the game.

Date Time

LESSON 7•3

Basketball Addition

	Points Scored			
	Team 1		Team 2	
	1st Half	2nd Half	1st Half	2nd Half
Player 1				
Player 2				
Player 3				
Player 4				
Player 5				
Team Score				

Point Totals	**1st Half**	**2nd Half**	**Final**
Team 1	______	______	______
Team 2	______	______	______

1. Which team won the first half? ________________

   By how much? ______ points

2. Which team won the second half? ________________

   By how much? ______ points

3. Which team won the game? ________________

   By how much? ______ points

Date Time

LESSON 7·3 More Multiplication Number Stories

Write your own multiplication stories and draw pictures of your stories. You can use the pictures at the side of the page for ideas.

For each story:

- Write the words.
- Draw a picture.
- Write the answer.

Example:

There are 5 tricycles. How many wheels in all?

Answer: 15 wheels
(unit)

1. ______________________________

____

Answer: ______________________
(unit)

2. ______________________________

____

Answer: ______________________
(unit)

A person has 2 ears.

A tricycle has 3 wheels.

A car has 4 wheels.

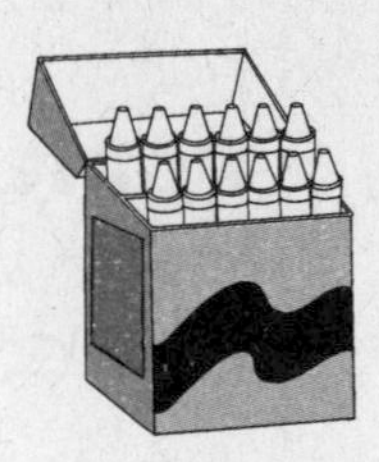

The box has 12 crayons.

The box has 100 paper clips.

The juice pack has 6 cans.

Date Time

LESSON 7·3

Math Boxes

1. Which one is impossible? Choose the best answer.

- ⬭ I will write a letter to a friend.
- ⬭ I will eat a piece of cake.
- ⬭ The sun will shine.
- ⬭ A fish will live out of water.

2. Solve.

Unit

_____ = 37 + 9

_____ = 137 + 9

116 − 8 = _____

176 − 8 = _____

3. Draw 5 fish bowls and 2 fish in each bowl.

How many fish in all?

_____ fish

4. Arrange the number of pets in order from the minimum (smallest) to the maximum (largest).

7, 0, 4, 1, 3, 5, 2

__, __, __, __, __, __, __

The maximum is _____.

The minimum is _____.

5. 1 bag of sugar weighs 5 pounds.

6 bags of sugar weigh

_____ pounds.

A. 50 **B.** 60

C. 25 **D.** 30

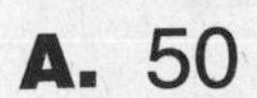

6. Circle the trapezoid that has $\frac{1}{3}$ shaded.

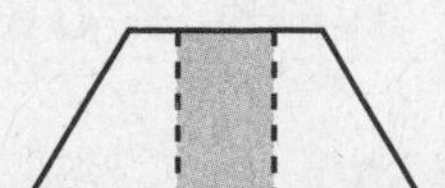

MRB 13

Date Time

LESSON 7·4

The Wubbles

1. On each line, write the number of Wubbles after doubling. Use your calculator to help you.

You started on Friday with _____ Wubble.

On Saturday, there were _____ Wubbles.

On Sunday, there were _____ Wubbles.

On Monday, there will be _____ Wubbles.

On Tuesday, there will be _____ Wubbles.

On Wednesday, there will be _____ Wubbles.

On Thursday, there will be _____ Wubbles.

On Friday, there will be _____ Wubbles.

A Wubble

2. On each line, write the number of Wubbles after halving. Use your calculator to help you. Remember that " $\frac{1}{2}$ of " means "divide by 2."

There were _____ Wubbles.

After Wink 1, there were _____ Wubbles.

After Wink 2, there were _____ Wubbles.

After Wink 3, there were _____ Wubbles.

After Wink 4, there were _____ Wubbles.

After Wink 5, there were _____ Wubbles.

After Wink 6, there were _____ Wubbles.

After Wink 7, there was _____ Wubble.

Your room could look like this! What will you do?

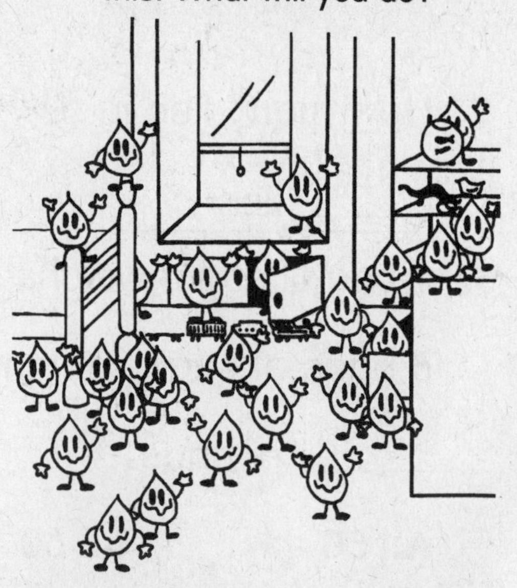

Adapted with permission from *Calculator Mathematics Book 2* by Sheila Sconiers, pp. 10 and 11 (Everyday Learning Corporation, © 1990 by the University of Chicago).

LESSON 7·4 Math Boxes

1. Collect 29 counters. How many groups of 3 can you make?

   ____ groups

   How many counters are left over?

   ____ counters

2. Solve.

   Unit
   train cars

   $4 + 3 + 13 =$ ____

   ____ $= 12 + 6 + 8$

   $5 + 4 + 18 =$ ____

   ____ $= 18 + 12 + 6$

   $40 = 15 + 6 +$ ____

3. Fill in the missing numbers.

	717	

4. Draw the lines of symmetry on this rectangle.

   How many lines of symmetry are there? ____

5. **Rule**
   Double

in	out
2	4
4	
5	
	14

6. Shade one half of this square.

Date Time

LESSON 7·5 Math Boxes

1. Draw hands to show 7:15.

2.

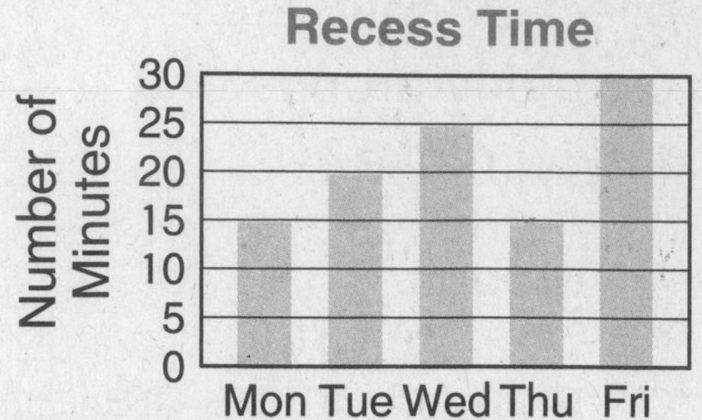

Which day had the longest recess? Circle the best answer.

A. Tuesday **B.** Friday

C. Wednesday **D.** Monday

3. Use counters to make a 5-by-3 array. Draw the array.

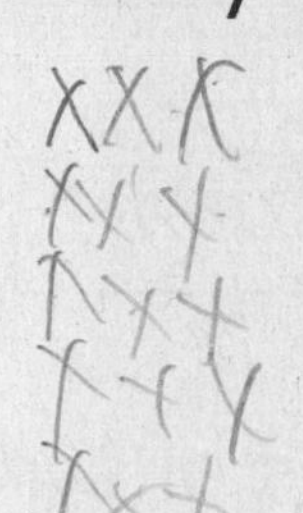

How many counters in all?

15 counters

4. Fill in the missing numbers.

608		
614		
	625	

5. Draw or write the names of two things in the classroom that are the shape of a rectangular prism.

6. Write the fraction.

The part shaded = $\frac{1}{4}$

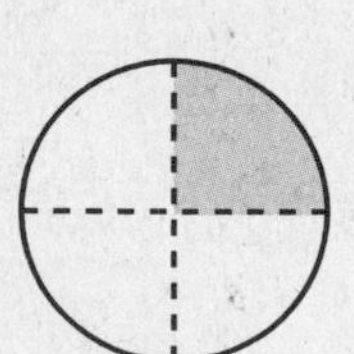

LESSON 7·6

Record of Our Jumps

You will make two jumps. For each jump, measure to the nearest centimeter and to the nearest inch.

How to Measure Each Jump

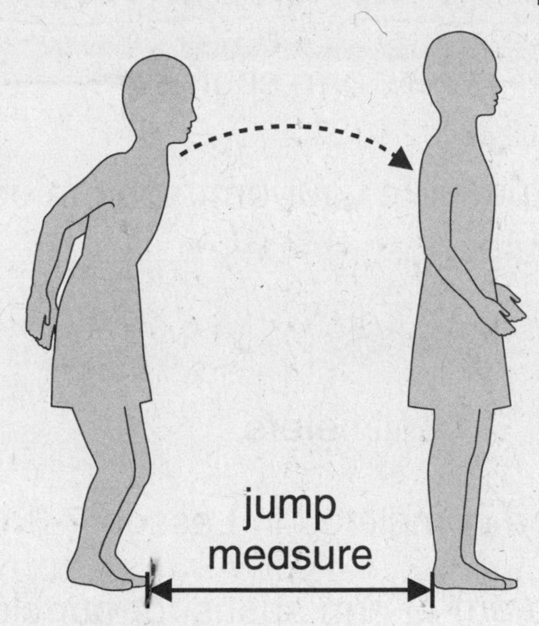

Place a penny or other marker (or make a dot with chalk) where the Jumper's back heel lands. Measure from the starting line to the marker.

1. Record your two jumps.

First jump:

_____ centimeters

_____ inches

Second jump:

_____ centimeters

_____ inches

2. Circle the measures for your longer jump.

You will complete the next question in Lesson 7-7.

3. A middle value of jumps for our class is _____ centimeters.

Date Time

Record of Our Arm Spans

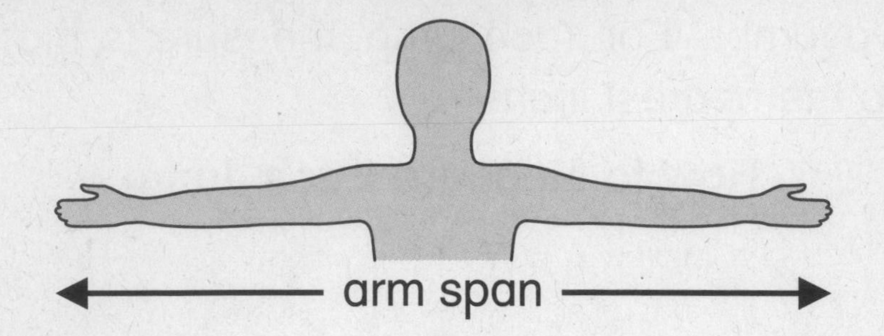

Work with a partner to measure your arm span in both inches and centimeters. Record your measures below.

1. My arm span is ______ inches.

2. My arm span is ______ centimeters.

The next question will be completed in Lesson 7-8.

3. A middle value (median) of arm spans for our class is ______ inches.

Date Time

LESSON 7·6

Math Boxes

1. Show five possible ways to make 40¢.

2. 2 bunches of bananas. Each bunch has 5 bananas. How many bananas in all?

_______ bananas

Complete the diagram.

bunches	bananas per bunch	bananas in all

3. What is the temperature? Fill in the circle next to the best answer.

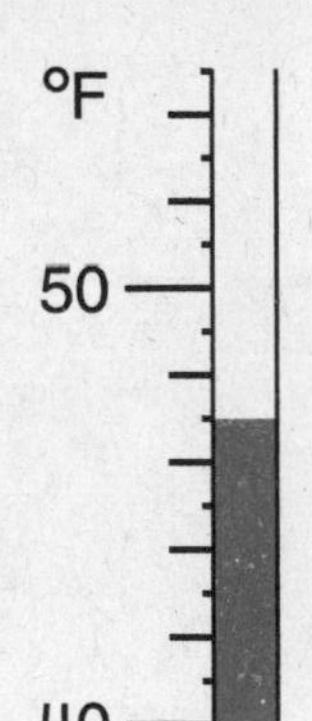

Ⓐ 43°

Ⓑ 53°

Ⓒ 46°

Ⓓ 47°

4. Which number is the most popular (the mode)?

Unit
minutes of homework

____

12, 12, 12, 14, 15, 16, 16

5. Make a ballpark estimate. Then solve the problem.
Ballpark estimate:

____

$$\begin{array}{r} 52 \\ -\ 29 \\ \hline \end{array}$$

6. Write <, >, or = in the box.

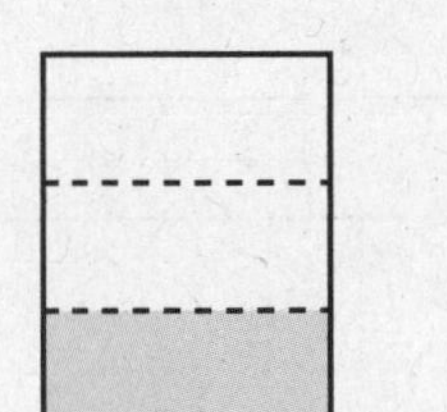

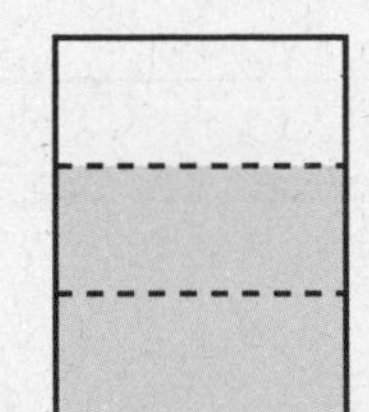

$\frac{1}{3}$  $\frac{2}{3}$

MRB 12–15

Date Time

LESSON 7•7

The Lengths of Objects

Reminder: *in.* means *inches; cm* means *centimeters*

Measure each item to the nearest inch.
Measure each item to the nearest centimeter.
Record your answers in the blank spaces.

1. pencil

about ______ in.

about ______ cm

2. screwdriver

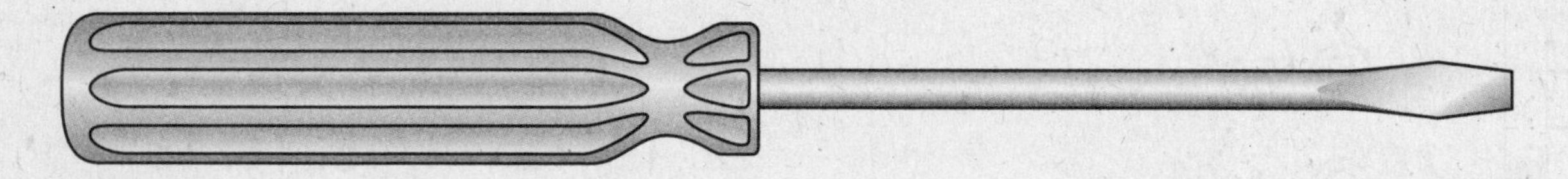

about ______ in.

about ______ cm

3. pen

about ______ in.

about ______ cm

LESSON 7•7

The Lengths of Objects *continued*

4. bolt

about ______ in.

about ______ cm

5. dandelion leaf

about ______ in.

about ______ cm

6. List the objects in order from shortest to longest.

____

____

____

____

____

7. How much longer is the pencil than the bolt?

______ in. ______ cm

8. How much longer is the screwdriver than the pen?

______ in. ______ cm

Date Time

LESSON 7·7

Math Boxes

1. Draw hands to show 4:15.

2. **Books Read**

Number of Books	Juan	Lilly	Grace	Terell
			X	
			X	
	X		X	
	X		X	X
	X		X	X
	X	X	X	X
	X	X	X	X

Who read the most books (maximum)? __________

Who read the least books (minimum)? __________

3. There are 6 rooms. Each room has 4 windows. How many windows in all? _____ windows

Draw an array.

4. Fill in the missing numbers.

892	
	903

5. Here is a picture of a pyramid. What is the shape of one of the faces? ______________

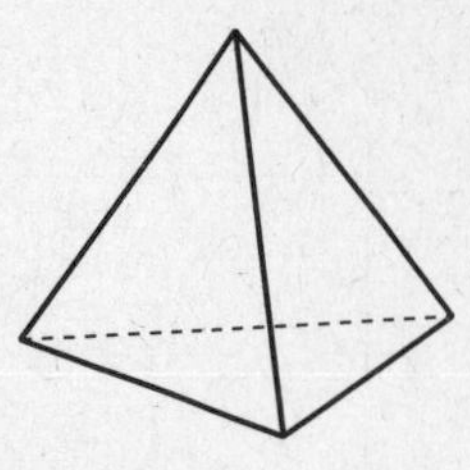

6. Color $\frac{5}{8}$ of the rectangle.

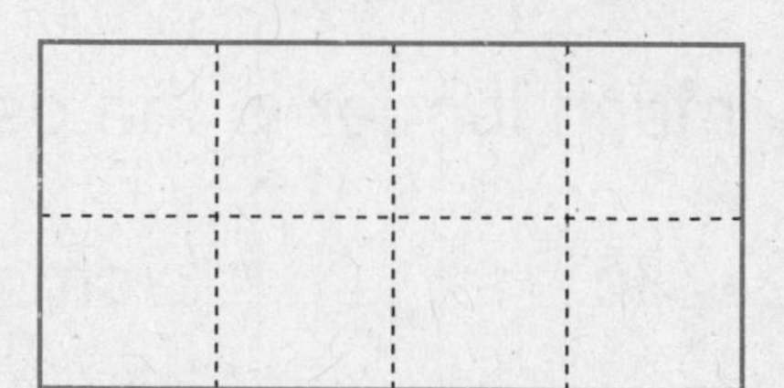

Date ____________ Time ____________

Soccer Spin Directions

Materials ☐ *Math Masters,* pp. 470 and 471

☐ counter

☐ paper clip

☐ pencil

Use a pencil and paper clip to make a spinner.

Players 2

Skill Predict outcomes of events

Object of the Game To test the prediction made at the beginning of the game

Directions

1. Players agree upon one spinner to use during the game.
2. Each player chooses a team to cheer for, **Checks** or **Stripes.** (Players can cheer for the same team.) They look at their spinner choice and predict which team will win the game.
3. The game begins with the counter in the center of the soccer field.
4. Players take turns spinning and moving the counter one space toward the goal that comes up on the spinner.
5. The game is over when the counter reaches a goal.
6. Players compare and discuss the results of their predictions. Play two more games using the other two spinners.

Follow-Up

1. Which spinner(s) would you want to use if you were cheering for the **Checks** team? Explain.
2. Which spinner(s) would you want to use if you were cheering for the **Stripes** team? Explain.

Date Time

Table of Our Arm Spans

Make a table of the arm spans of your classmates.

Our Arm Spans		
Arm Span (inches)	**Frequency**	
	Tallies	**Number**
	Total =	

Date Time

Bar Graph of Our Arm Spans

Make a bar graph of the arm spans of your classmates.

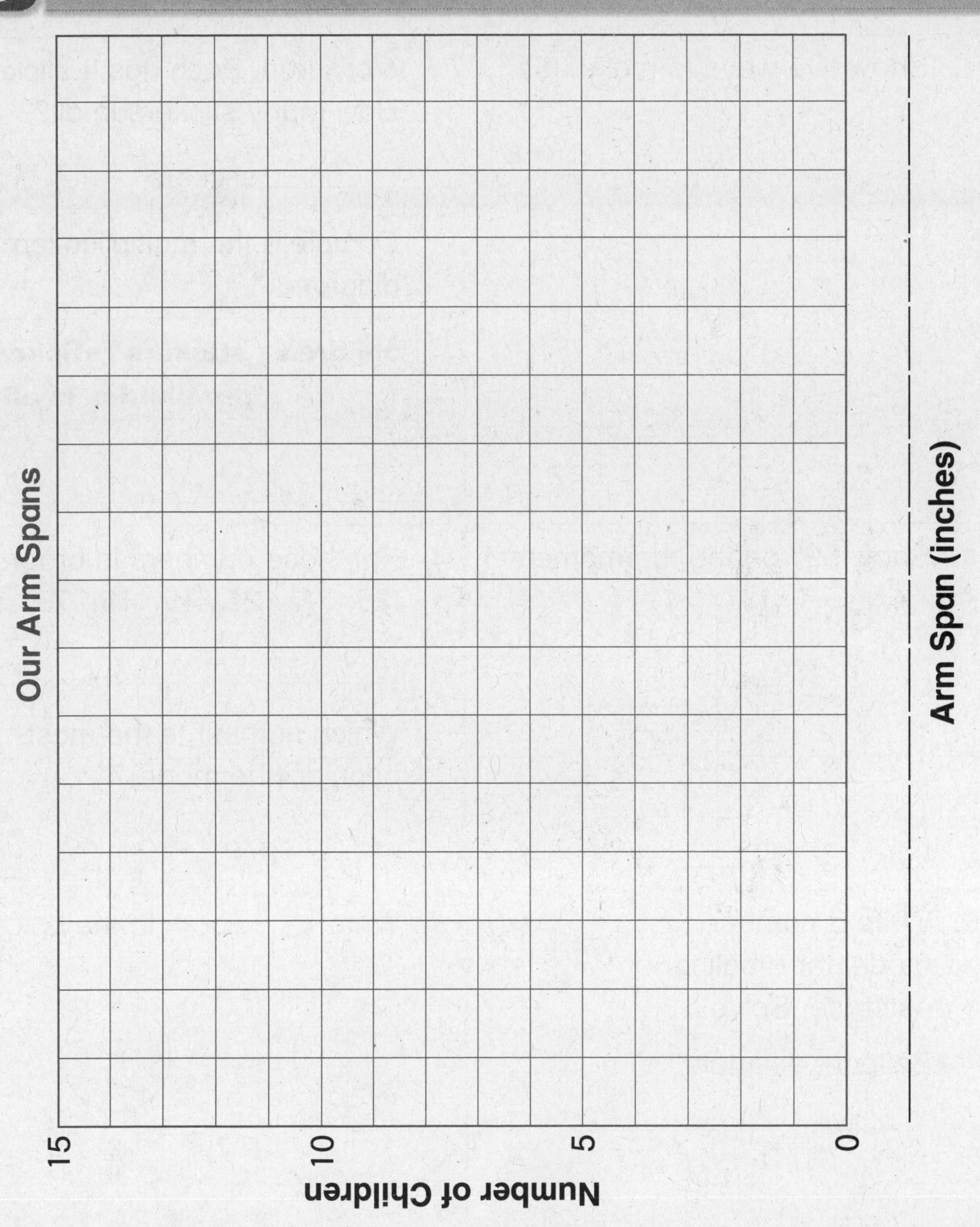

Date Time

Math Boxes

1. Show five ways to make 45¢.

2. 6 children. Each has 4 stickers. How many stickers in all?

_______ stickers

Complete the multiplication diagram.

children	stickers per child	stickers in all

3. Show 52° on the thermometer.

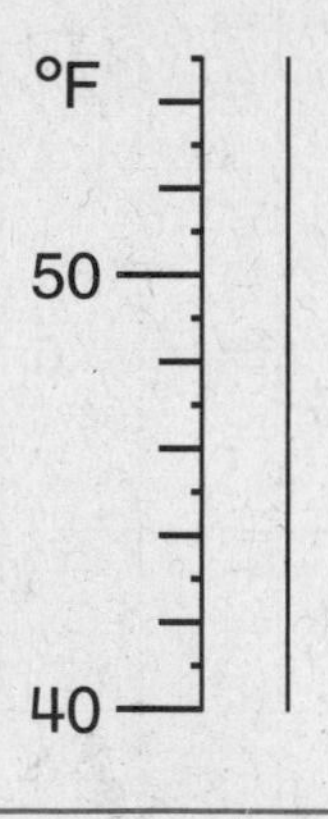

4. Put these numbers in order.
25, 15, 25, 19, 15, 75, 15

____

Which number is the most popular (the mode)?

____

5. Write a number model for a ballpark estimate. Solve.

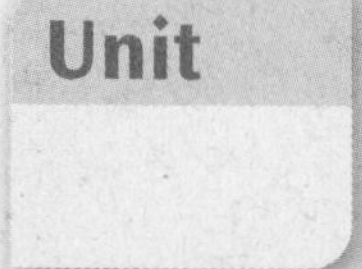

Ballpark estimate:

____

$$\begin{array}{r} 47 \\ -\ 31 \\ \hline \end{array}$$

6. Write <, >, or = in the box.

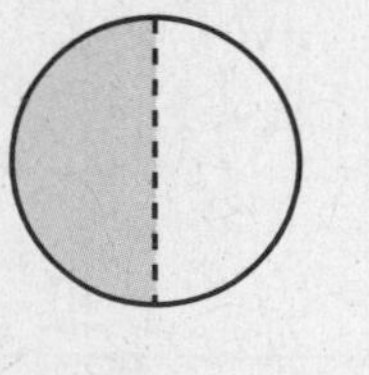
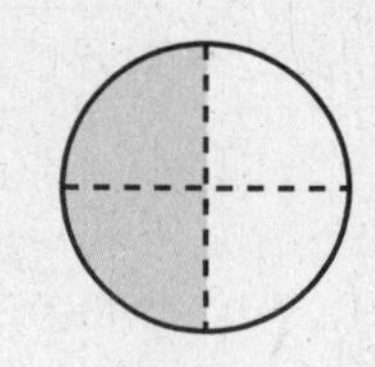

$\frac{1}{2}$  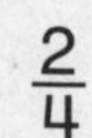 $\frac{2}{4}$

Date Time

Math Boxes

1. 

What fraction of the triangles are shaded?

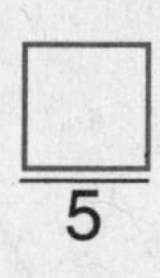

$\frac{\square}{5}$

2. Fill in the box.

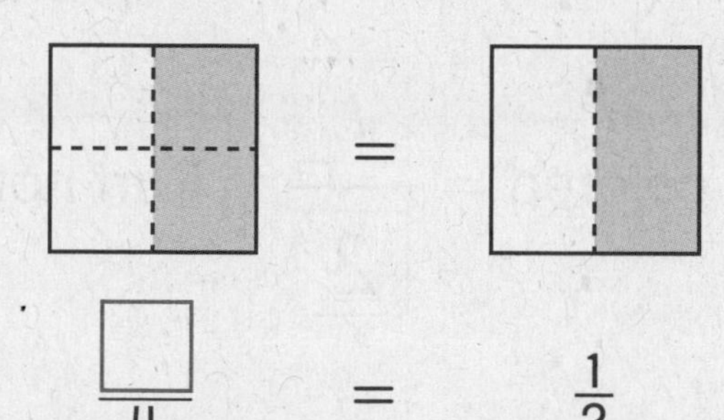

$\frac{\square}{4} = \frac{1}{2}$

3. Hana has 6 bracelets to wear. 3 of them are made of beads. What fraction of the bracelets are made of beads?

______

4. Shade $\frac{1}{3}$ of this trapezoid.

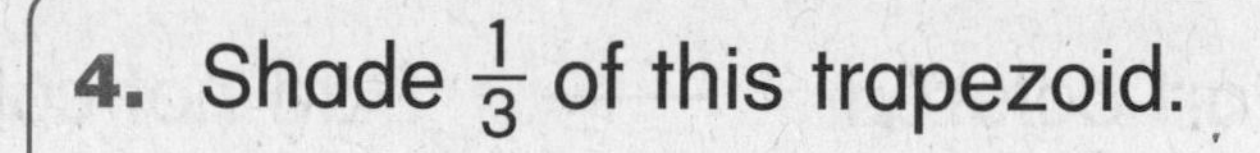

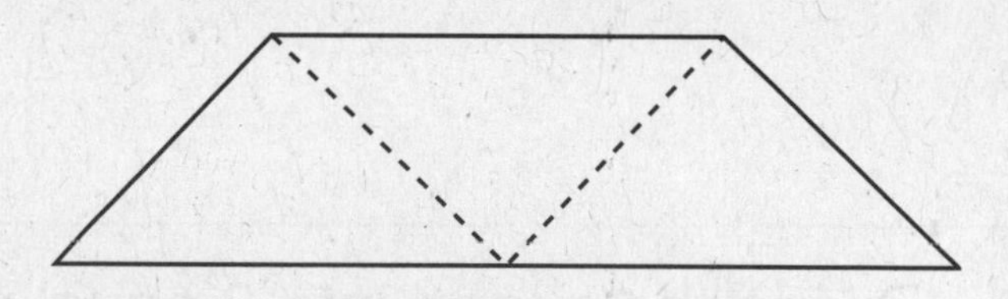

5. Circle $\frac{1}{4}$ of the pennies.

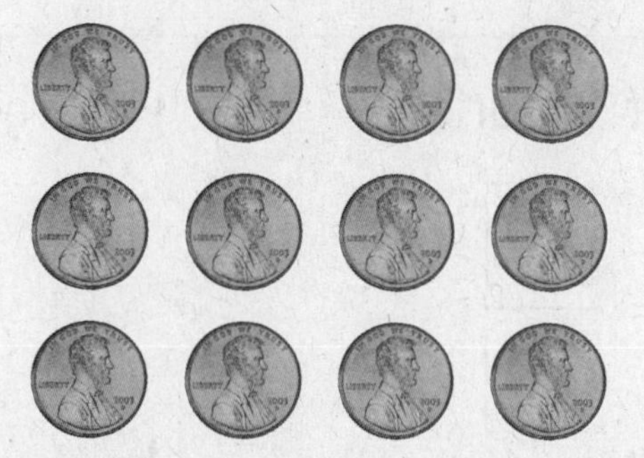

6. If 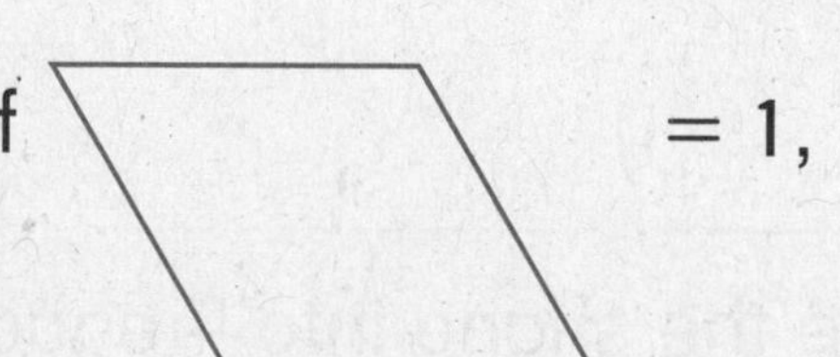 = 1,

then  = ______

Date Time

LESSON 8·1 Equal Parts

Use a straightedge or Pattern-Block Template.

1. Divide the shape into 2 equal parts. Color 1 part.

Part colored = $\frac{1}{2}$ Part not colored = $\frac{\square}{\square}$

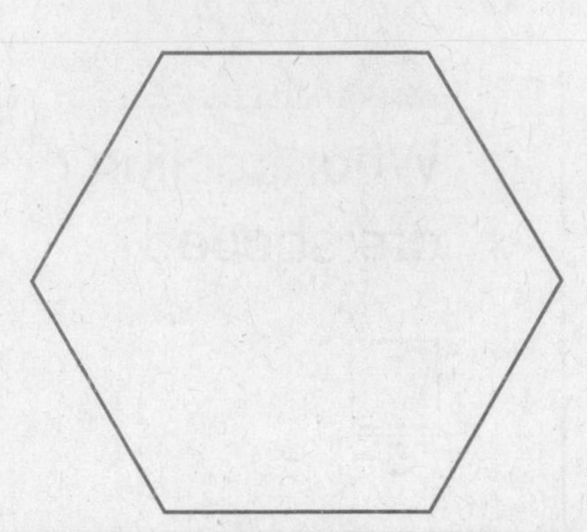

2. Divide the shape into 6 equal parts. Color 1 part.

Part colored = $\frac{\square}{\square}$ Part not colored = $\frac{\square}{\square}$

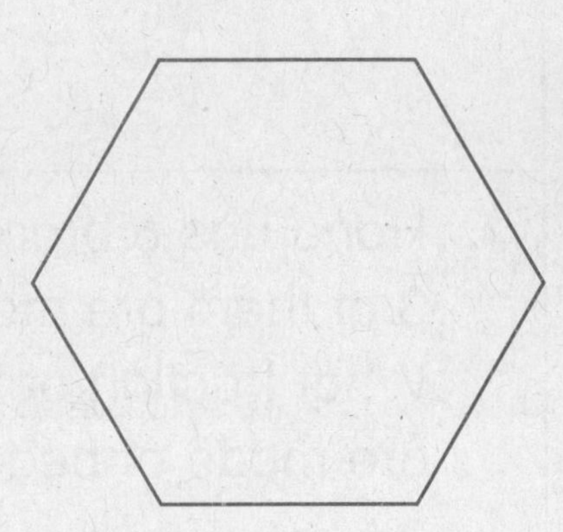

3. Divide the shape into 3 equal parts. Color 2 parts.

Part colored = $\frac{\square}{\square}$ Part not colored = $\frac{\square}{\square}$

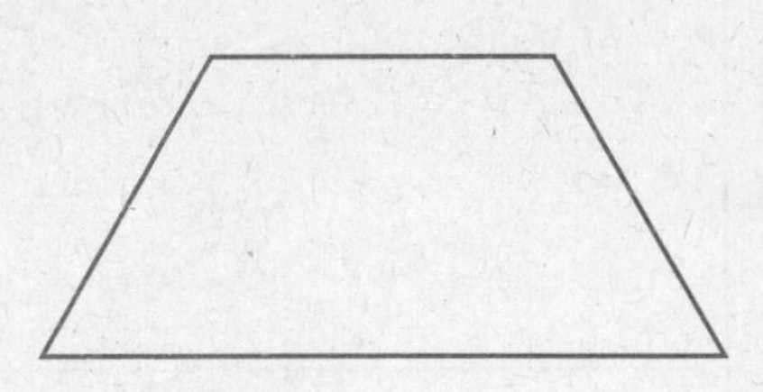

4. Divide the shape into 4 equal parts. Color 2 parts.

Part colored = $\frac{\square}{\square}$ Part not colored = $\frac{\square}{\square}$

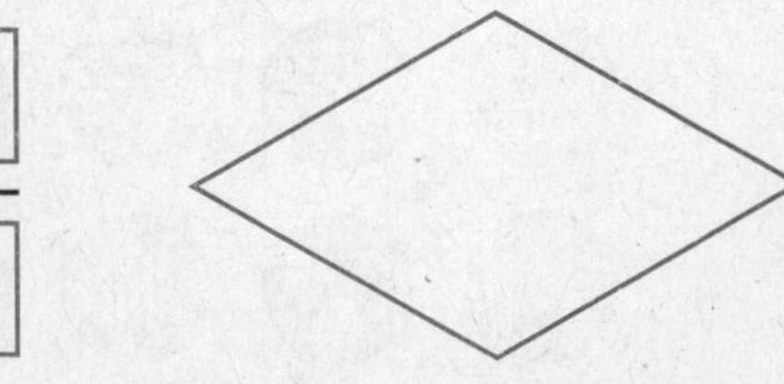

Date ______ Time ______

LESSON 8•1

Bamboo Plant Number Stories

Bamboo is one of the world's fastest-growing plants. Some types of bamboo grow more than 24 inches per day and reach heights close to 100 feet! For one week, a growing bamboo plant was measured. The chart below shows its height for each day.

Bamboo Plant Growth for One Week

Sun.	Mon.	Tues.	Wed.	Thurs.	Fri.	Sat.
12 in.	26 in.	40 in.	57 in.	63 in.	80 in.	99 in.

Use the information above to solve the following number stories.

1. How many inches did the bamboo plant grow from Tuesday to Friday?

   Answer: ______ inches

   Number model:

   ____

2. How many inches did the bamboo plant grow from Thursday to Friday?

   Answer: ______ inches

   Number model:

   ____

3. How many more inches tall was the bamboo plant on Saturday than on Tuesday?

   Answer: ______ inches

   Number model:

   ____

4. How many more inches tall was the bamboo plant on Saturday than Sunday?

   Answer: ______ inches

   Number model:

   ____

Date Time

LESSON 8·1 Math Boxes

1. Write fractions.

   The part shaded = ______.

   The part not shaded = ______.

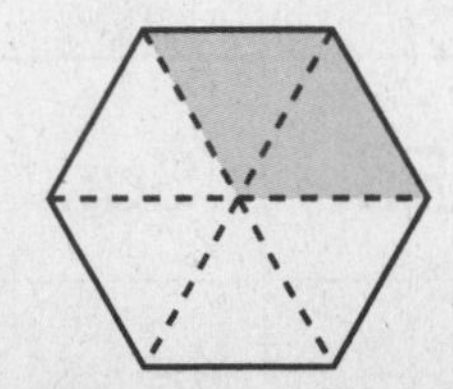

2. Use a Pattern-Block Template. Draw a shape that has at least one line of symmetry.

3. Complete the table.

in	out
2	1
4	
8	
	5

Rule
$\frac{1}{2}$

4. Show 2 ways to make 50¢. Use Ⓠ, Ⓓ, Ⓝ, and Ⓟ.

5. Circle the thing you are certain will happen.

   You will roll a 7 on a die.

   The temperature will be exactly 20°F today.

   An hour will pass.

6. grapes watermelon

   Which item is heavier?

   ____

Date Time

LESSON 8·2

Pattern-Block Fractions

Use pattern blocks to help you solve each problem.

Use your Pattern-Block Template to show what you did.

Example:

If [trapezoid] = 1, then  = $\frac{1}{3}$.

1. If [shape] = 1, then 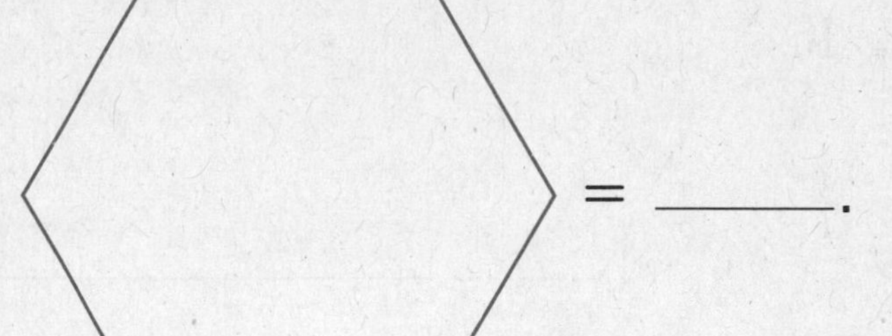 = ______.

2. If  [large triangle] = 1, then  = ______.

Date   Time

LESSON 8·2

Pattern-Block Fractions *continued*

3. If  = 1, then 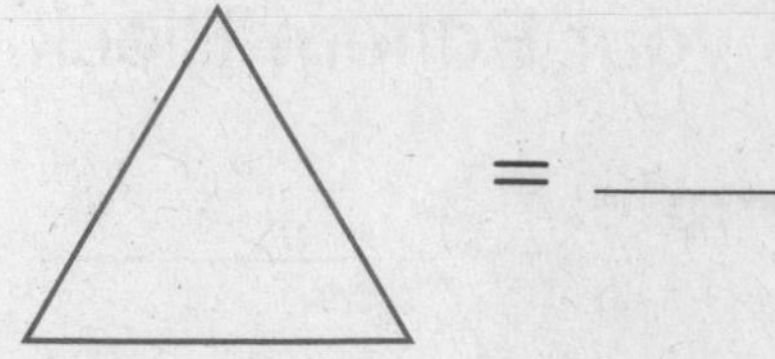 = ______.

4. If = 1, then 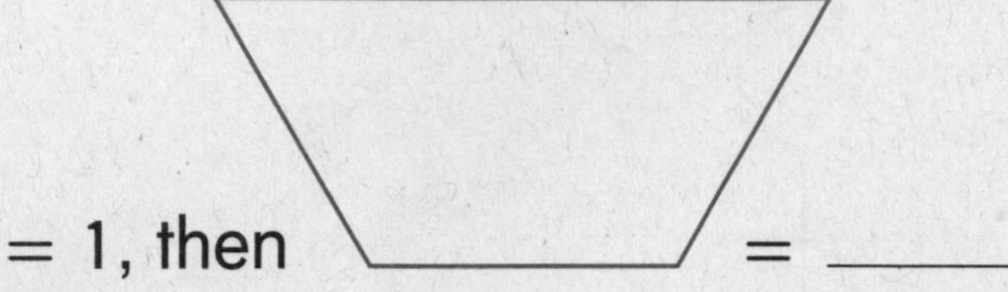 = ______.

5. If = 1, then 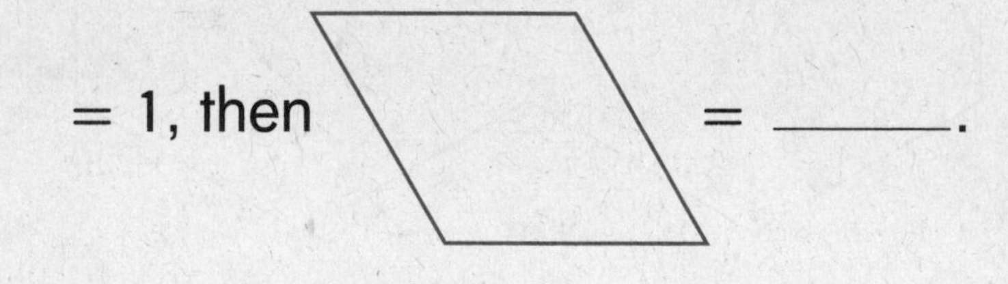 = ______.

6. If = 1, then 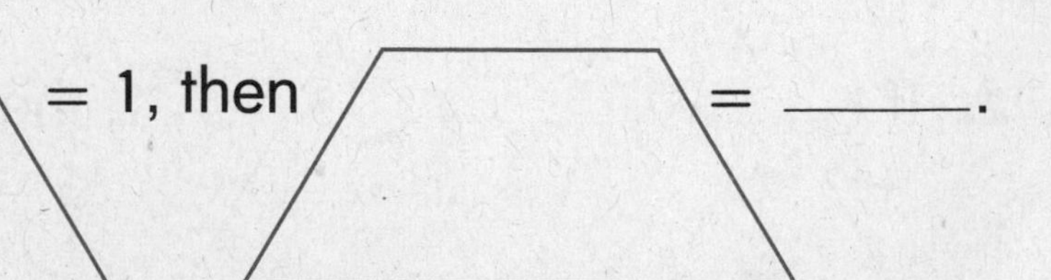 = ______.

Date Time

LESSON 8•2

Geoboard Fences

1.

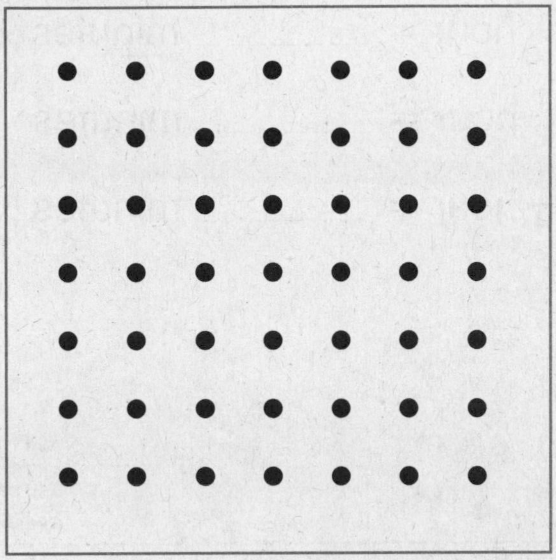

2.

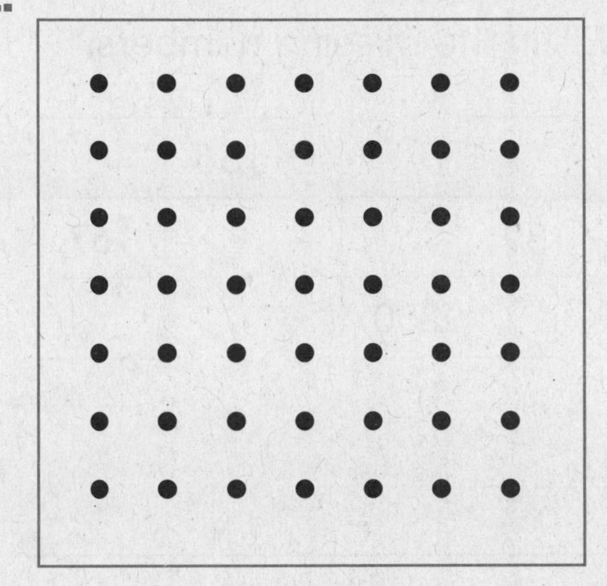

3.

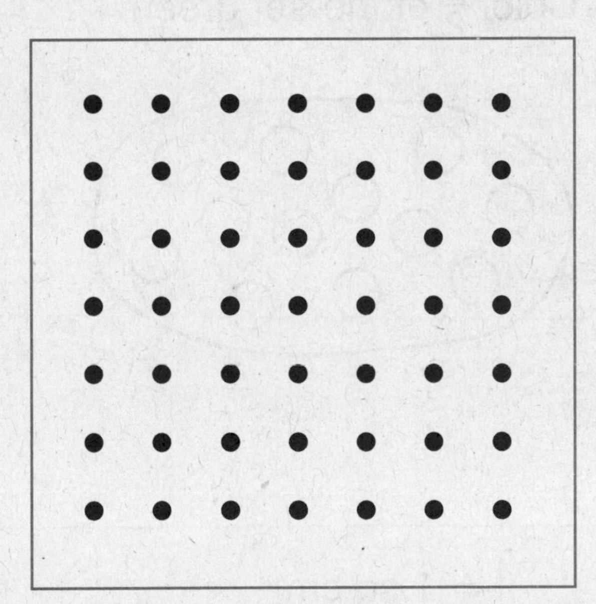

4.

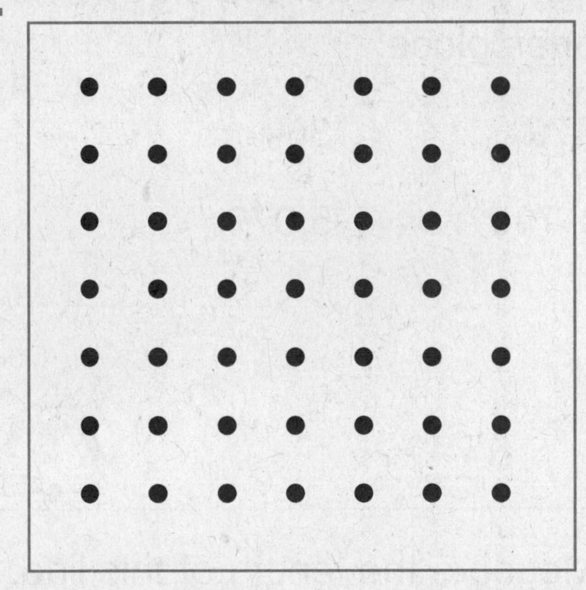

Fence	How many pegs in all?	How many rows of pegs?	How many in each row?
1.			
2.			
3.			
4.			

LESSON 8•2

Math Boxes

1. Fill in the missing numbers.

189	
	200

156	
	167

2. 1 hour = ________ minutes

$\frac{1}{2}$ hour = ________ minutes

$\frac{1}{4}$ hour = ________ minutes

3. Put a line under the digit in the ones place.

479 364

1,796 5,079

4. Color $\frac{1}{2}$ of the set green.

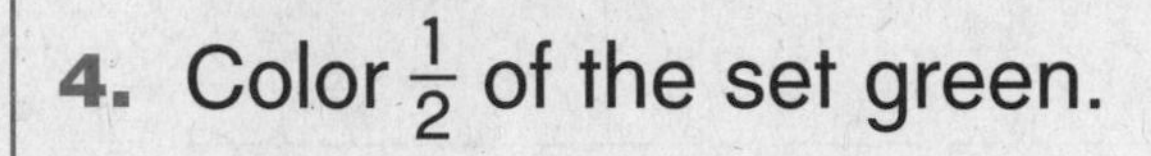

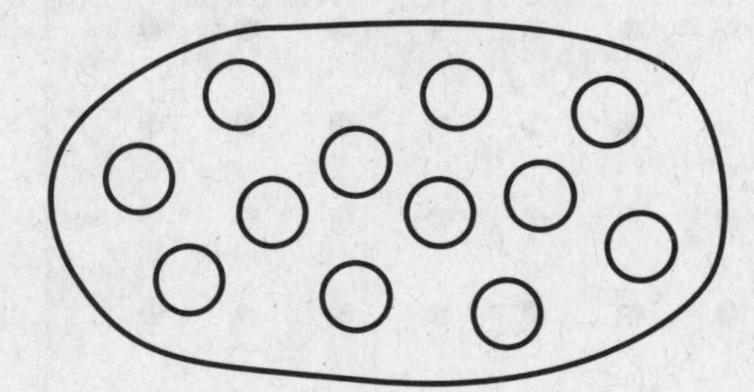

5. Measure the length of this line.

about ________ cm

about ________ in.

6. ☐ = 1 sq cm

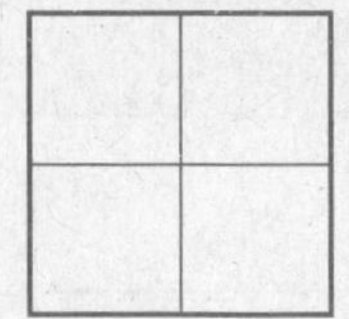

Area = ________ sq cm

Date Time

LESSON 8·3

Equal Shares

Use pennies to help you solve the problems.

Circle each person's share.

1. Two people share 10 pennies. How many pennies does each person get?

   _____ pennies

   $\frac{1}{2}$ of 10 pennies = _____ pennies.

2. Three people share 9 pennies. How many pennies does each person get?

   _____ pennies

   $\frac{1}{3}$ of 9 pennies = _____ pennies.

   $\frac{2}{3}$ of 9 pennies = _____ pennies.

3. Four people share 12 pennies. How many pennies does each person get?

   _____ pennies

   $\frac{1}{4}$ of 12 pennies = _____ pennies.

   $\frac{3}{4}$ of 12 pennies = _____ pennies.

Date ______________________ Time ______________________

LESSON 8·3

Fractions of Sets

A fraction is given in each problem. Color that fraction of the checkers red.

1. $\frac{1}{5}$ are red.	**2.** $\frac{2}{3}$ are red.
3. $\frac{3}{4}$ are red.	**4.** $\frac{4}{6}$ are red.
5. $\frac{1}{2}$ are red.	**6.** $\frac{0}{7}$ are red.

Try This

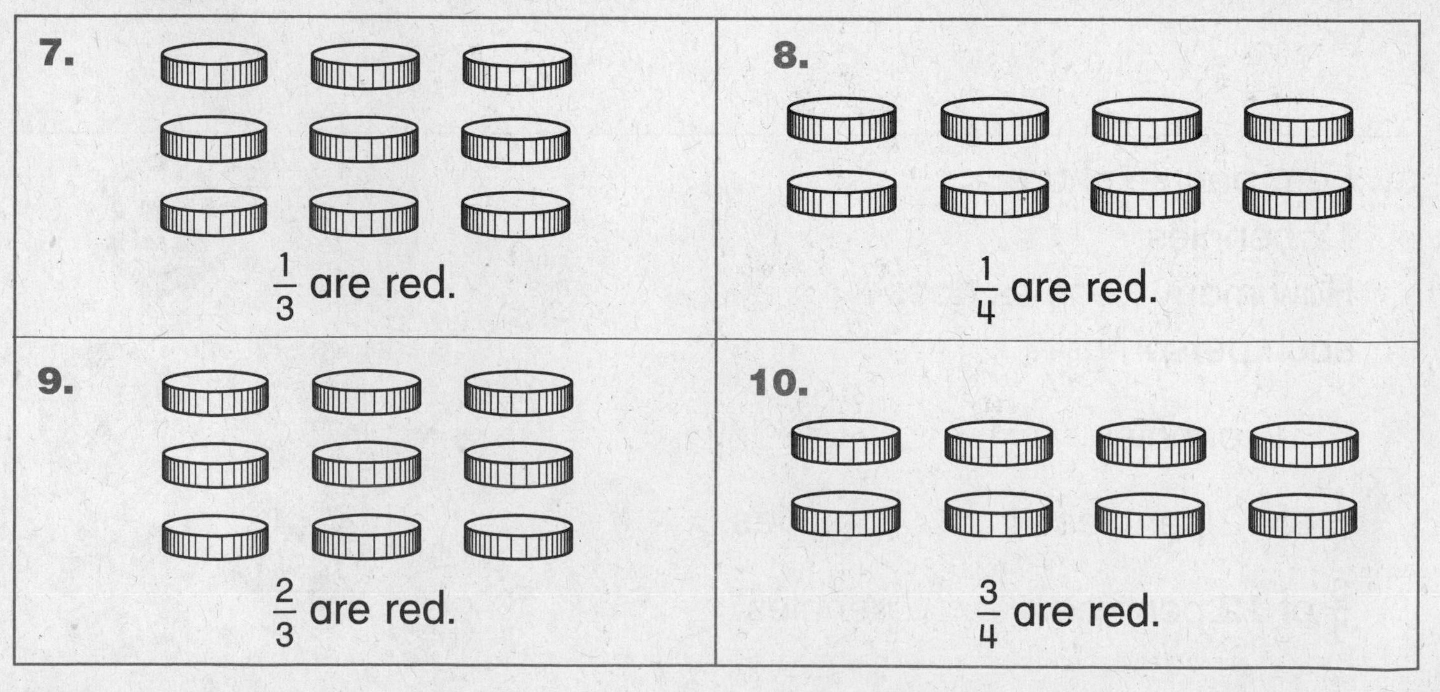

Date Time

LESSON 8·3

Equal Parts

Use a straightedge or Pattern-Block Template.

1. Divide the shape in half.

How many halves? ________

2. Divide the shape into 4 equal parts.

How many fourths? ________

3. Divide the shape in half.

How many halves? ________

4. Divide the shape into 4 equal parts.

How many fourths? ________

Try This

5. Use your Pattern-Block Template triangle.
Divide the shape into 3 equal parts.

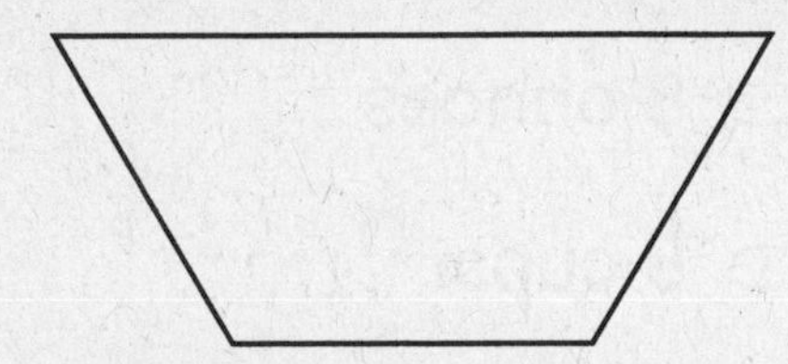

How many thirds? ________

LESSON 8•3 Math Boxes

1. Color $\frac{1}{4}$ blue. Color $\frac{1}{4}$ yellow. Color $\frac{1}{2}$ red.

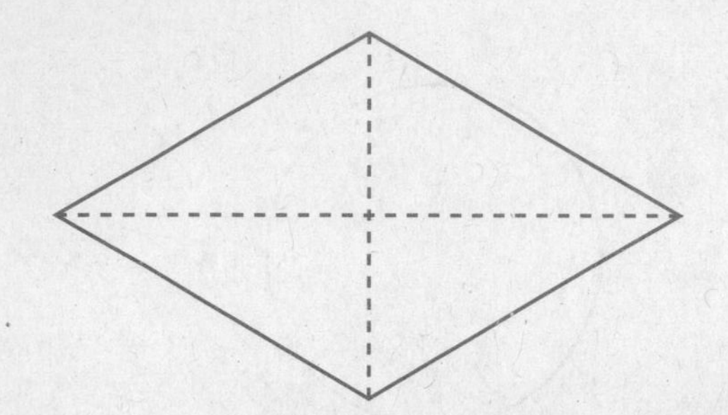

2. Circle the figure that has only one line of symmetry. Draw the line of symmetry.

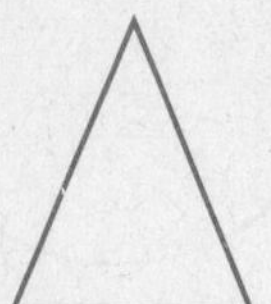

3. Complete the table.

Rule
1 yd = 3 ft

yd	ft
2	
	9
5	
	30

4. Show 1 way to make $1.28. Use Ⓠ, Ⓓ, Ⓝ, and Ⓟ.

5. Circle the event that is likely to happen.

You will fly to the center of the earth.

You will have homework.

You will eat a rock.

6. Which unit makes sense? Choose the best answer.

A can of soup may weigh:

- ⬭ 8 ounces
- ⬭ 8 cups
- ⬭ 8 pounds
- ⬭ 8 feet

LESSON 8•4

Math Boxes

1. Fill in the missing numbers.

	1,217	

2. ______ months = 1 year

______ months = $\frac{1}{2}$ year

______ months = $\frac{1}{4}$ year

______ months = 2 years

3. Circle the digits in the hundreds place.

1 2 8 9 7 2 4 6 3

2, 4 6 5 3, 0 9 1

6 6, 2 5 0

4. There are 9 dinosaurs. 3 are plant eaters. Which fraction shows how many are plant eaters? Fill in the circle next to the best answer.

Ⓐ $\frac{9}{3}$ Ⓒ $\frac{1}{2}$

Ⓑ $\frac{3}{9}$ Ⓓ $\frac{2}{3}$

5. Draw a triangle. Measure each side to the nearest inch.

about ___ in.

about ___ in.

about ___ in.

6. Count the shaded squares to find the area.

Area = ___ sq cm

Date Time

LESSON 8·4

Equivalent Fractions

Do the following:

- Use the circles that you cut out of *Math Masters,* page 239.
- Cut these circles apart along the dashed lines.
- Glue the cutout pieces onto the circles on this page and the next, as directed.
- Write the missing numerators to complete the equivalent fractions.

1. Cover $\frac{1}{2}$ of the circle with fourths.

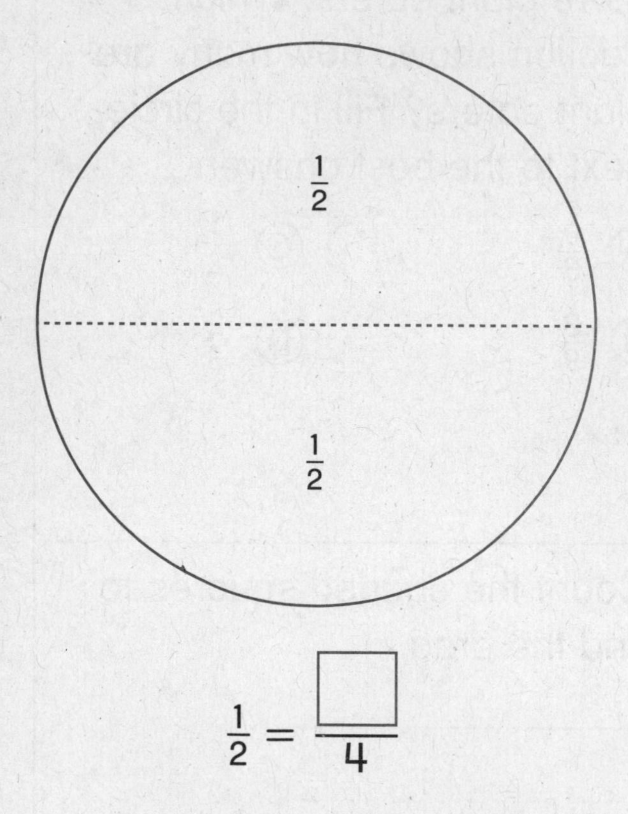

$\frac{1}{2} = \frac{\square}{4}$

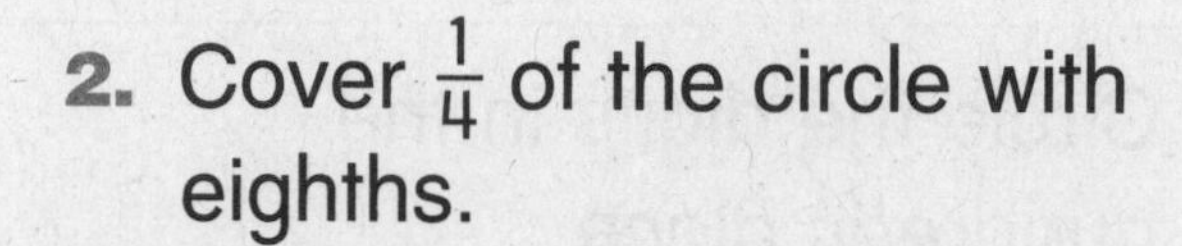

2. Cover $\frac{1}{4}$ of the circle with eighths.

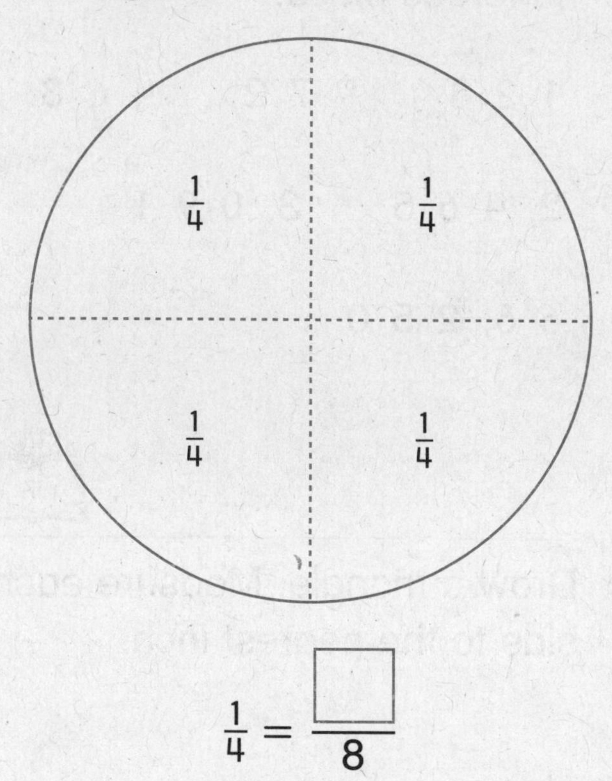

$\frac{1}{4} = \frac{\square}{8}$

Date _______________ Time _______________

LESSON 8•4

Equivalent Fractions *continued*

3. Cover $\frac{2}{4}$ of the circle with eighths.

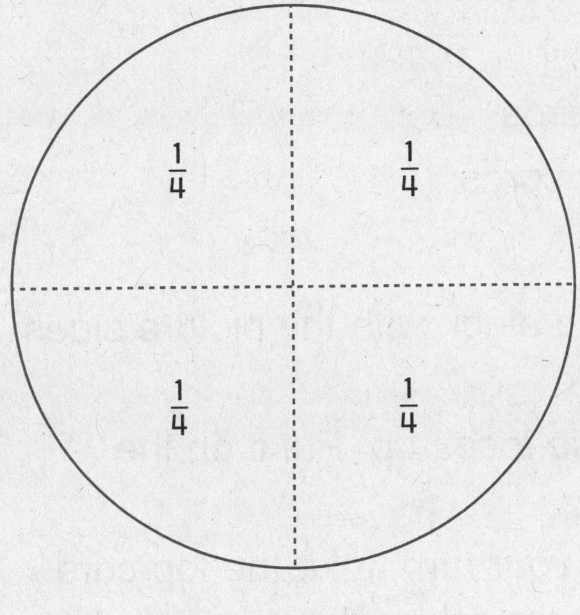

$\frac{2}{4} = \frac{\square}{8}$

4. Cover $\frac{1}{2}$ of the circle with sixths.

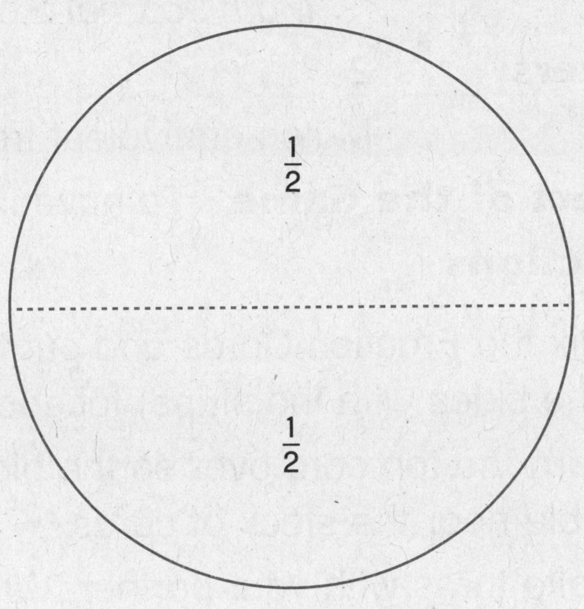

$\frac{1}{2} = \frac{\square}{6}$

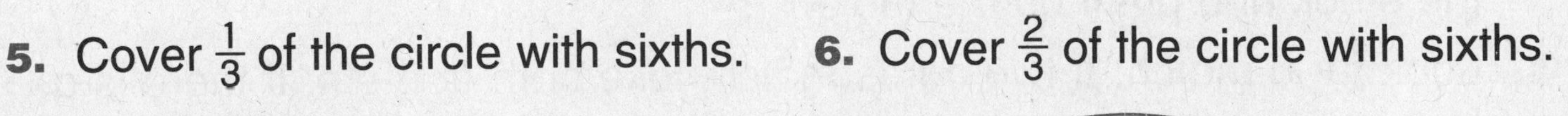

5. Cover $\frac{1}{3}$ of the circle with sixths.

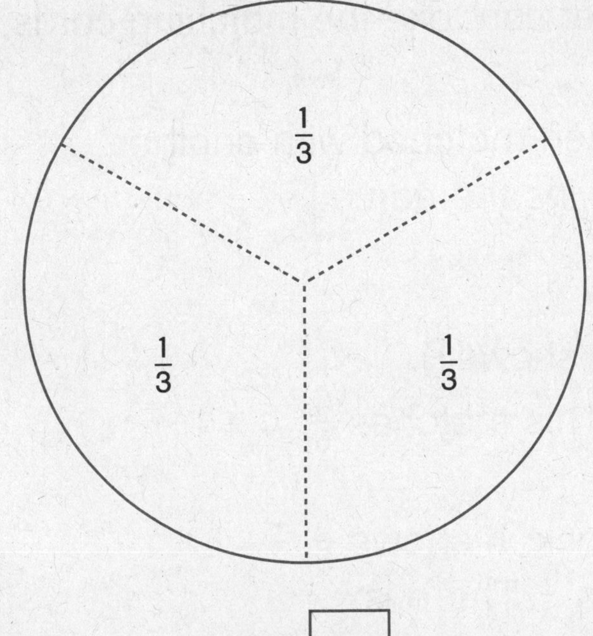

$\frac{1}{3} = \frac{\square}{6}$

6. Cover $\frac{2}{3}$ of the circle with sixths.

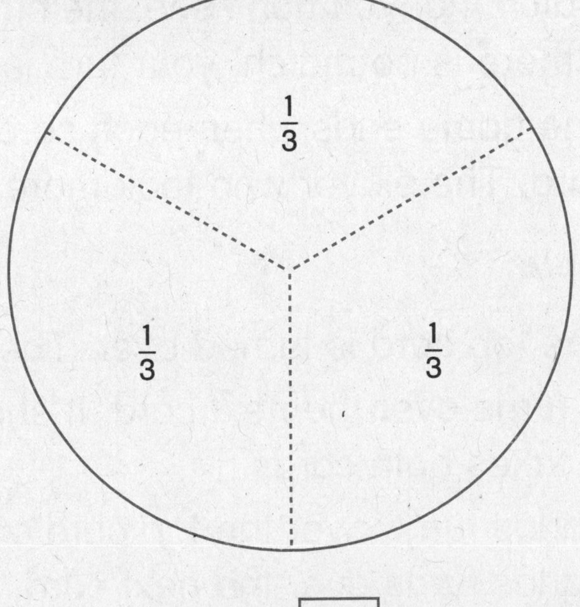

$\frac{2}{3} = \frac{\square}{6}$

LESSON 8•5

Equivalent Fractions Game Directions

Materials ☐ 32 Fraction Cards (2 sets cut from *Math Journal 2,* Activity Sheets 5 and 6)

Players 2

Skill Match equivalent fraction cards

Object of the Game To have the most cards

Directions

1. Mix the Fraction Cards and put them in a stack with the picture sides (the sides with the strips) facedown.
2. Turn the top card over so the picture side faces up. Put it on the table near the stack of cards.
3. Take turns with your partner. When it is your turn, take the top card from the stack. Turn it over and put it on the table. Try to match this card with a picture-side-up card on the table. (If there are no other picture-side-up cards on the table, turn over the next card on the stack and put it on the table.)
4. Look for a match. If two cards match, take both of them. If there is a match that you don't see, the other player can take the matching cards. If there is no match, your turn is over.
5. The game ends when each card has been matched with another card. The player who took more cards wins the game.

Example:

1. The top card is turned over. The picture shows $\frac{4}{6}$.
2. Li turns over the next card. It shows $\frac{2}{3}$. This matches $\frac{4}{6}$. Li takes both cards.
3. Carlos turns over the top card on the stack. It shows $\frac{6}{8}$. Carlos turns over the next card. It shows $\frac{0}{4}$. There is no match. Carlos places $\frac{0}{4}$ next to $\frac{6}{8}$.
4. Li turns over the top card on the stack.

LESSON 8·5

Equivalent Fractions Game Directions *continued*

Another Version

1. Mix the Fraction Cards and put them in a stack with the picture sides facedown.
2. Turn the top card over so the picture side faces up. Put it on the table with the picture side faceup.
3. Players take turns. When it is your turn, take the top card from the stack, but do *not* turn it over. Keep the picture side down. Try to match the fraction on the card with one of the picture-side-up cards on the table.
4. If you find a match, turn your card over. Check that your match is correct by comparing the two pictures. If your match is correct, take both cards.

   If there is no match, place your card next to the other cards, picture side faceup. Your turn is over. If the other player can find a match, he or she can take the matching cards.
5. If there are no picture cards showing when Player 2 begins his or her turn, take the top card from the stack. Place it on the table with the picture side showing. Then Player 2 takes the next card in the stack and doesn't turn that card over.

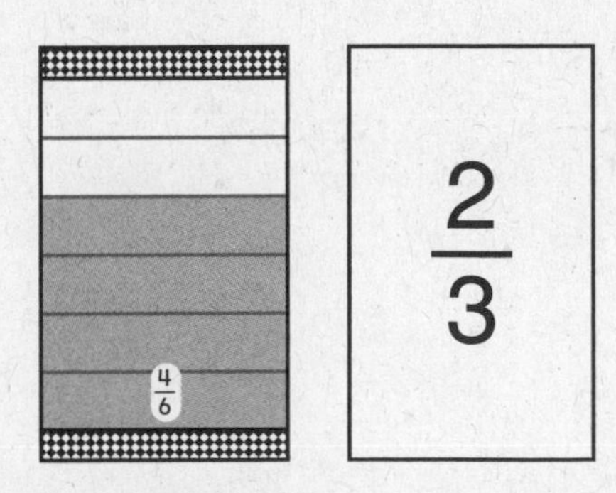

Marta thinks these two cards are a matching pair.

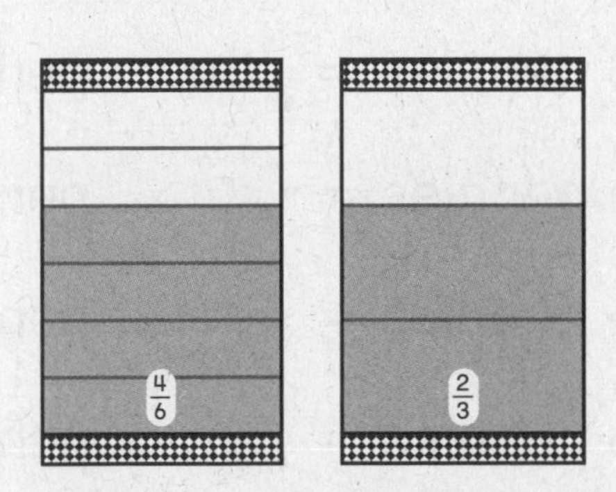

Marta checks that her match is correct by comparing the pictures of the fractions.

LESSON 8•5

Fractions of Collections

Use pennies to help you solve the problems.

1. Five people share 15 pennies.

How many pennies does each person get? _____ pennies

$\frac{1}{5}$ of 15 pennies = _____ pennies.

$\frac{2}{5}$ of 15 pennies = _____ pennies.

2. Six people share 12 pennies.

How many pennies does each person get? _____ pennies

$\frac{1}{6}$ of 12 pennies = _____ pennies.

$\frac{4}{6}$ of 12 pennies = _____ pennies.

3. Four people share 16 pennies.

How many pennies does each person get? _____ pennies

$\frac{1}{4}$ of 16 pennies = _____ pennies.

$\frac{4}{4}$ of 16 pennies = _____ pennies.

$\frac{2}{4}$ of 16 pennies = _____ pennies.

$\frac{3}{4}$ of 16 pennies = _____ pennies.

$\frac{0}{4}$ of 16 pennies = _____ pennies.

Date Time

LESSON 8·5

Fractions of Collections *continued*

Color the fractions of circles blue.

4. $\frac{3}{5}$ are blue.

5. $\frac{1}{2}$ are blue.

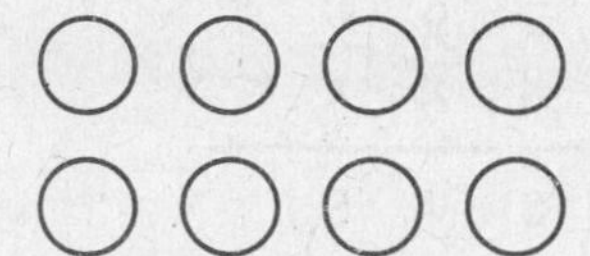

6. $\frac{1}{3}$ are blue.

7. $\frac{2}{3}$ are blue.

8. $\frac{3}{5}$ are blue.

9. $\frac{3}{4}$ are blue.

Try This

10. $\frac{3}{8}$ are blue.

11. $\frac{2}{6}$ are blue.

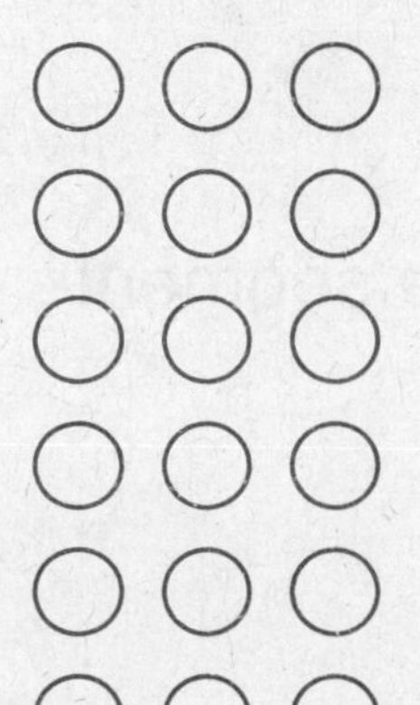

LESSON 8•5

Math Boxes

1. **Scores on a 5-Word Spelling Test**

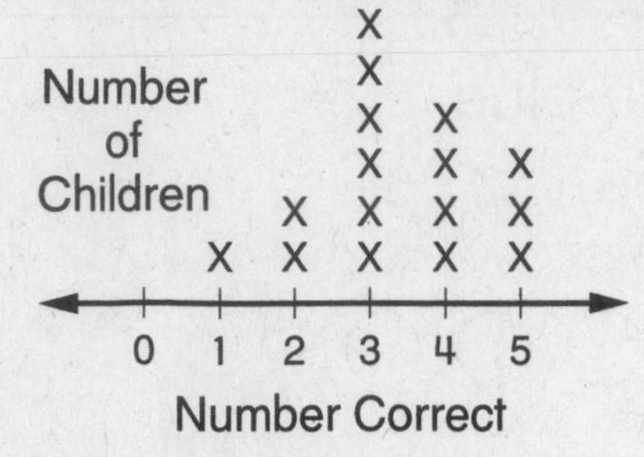

What score did the most children get (the *mode*)? _____

2. Circle $\frac{1}{5}$ of the nickels.

3. Find the arrow rules.

4. Complete the diagram.

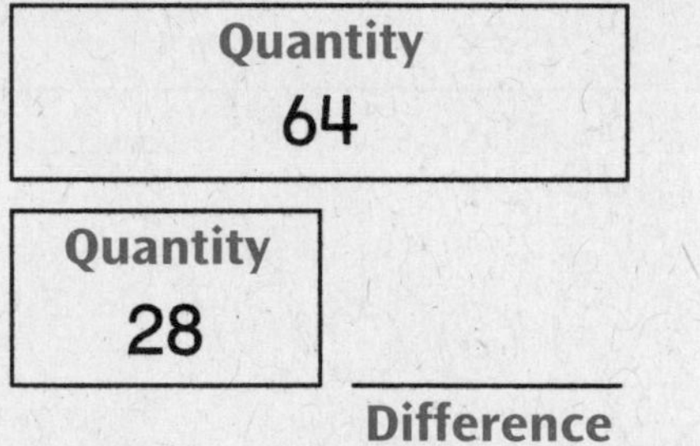

Write a number model.

____

5. Draw a line segment $4\frac{1}{2}$ cm long.

Now draw a line segment 2 cm longer.

6. Measure each side of the triangle to the nearest inch.
Find the perimeter.

The perimeter is

_____ inches.

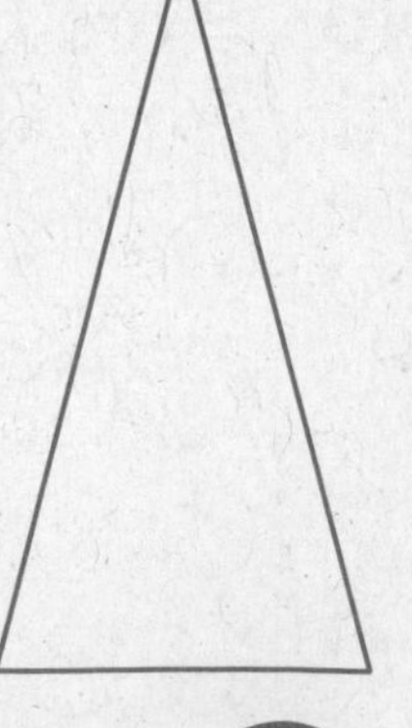

Date Time

LESSON 8·6

Fraction Top-It Directions

Use your Fraction Cards. List all the fractions that are:

less than $\frac{1}{2}$. ______________________

more than $\frac{1}{2}$. ______________________

the same as $\frac{1}{2}$. ______________________

Fraction Top-It

Materials ☐ 32 Fraction Cards (2 sets cut from *Math Journal 2,* Activity Sheets 5 and 6)

Players 2

Skill Compare fractions

Object of the Game To have the most cards

Directions

1. Mix the Fraction Cards and put them in a stack so all the picture sides (the sides with the strips) are facedown.
2. Each player turns over a card from the top of the stack. Players compare the shaded parts of their cards. The player with the larger (higher) fraction takes both cards.
3. If the shaded parts are equal, the fractions are equivalent. Each player turns over another card. The player with the larger fraction takes all the cards from both plays.
4. The game ends when all the cards have been taken from the stack. The player who took more cards wins.

$\frac{1}{2}$ $\frac{1}{2}$ is greater than $\frac{1}{3}$. $\frac{1}{3}$

Fraction Top-It Directions *continued*

Another Version

1. Mix the Fraction Cards and put them in a stack so all the picture sides (the sides with the strips) are facedown.

2. Each player takes a card from the top of the stack but does *not* turn it over.

3. Players take turns. When it is your turn, compare the fractions on the two cards. Say one of the following:

   - My fraction is more than your fraction.
   - My fraction is less than your fraction.
   - The fractions are equivalent.

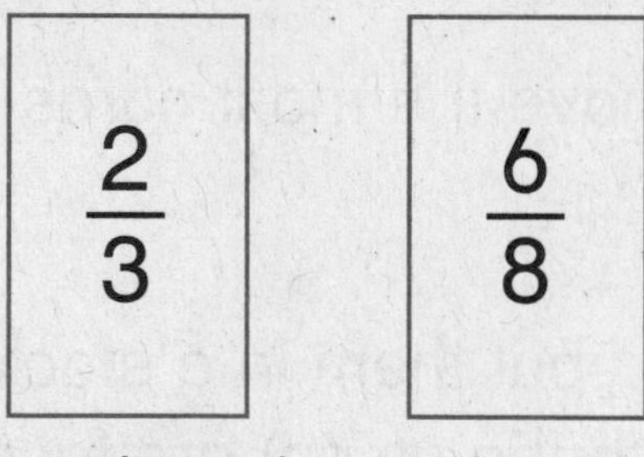

Denzel's card Barb's card

Denzel says that his fraction is less than Barb's fraction.

4. Turn the cards over and compare the shaded parts. If you were correct, take both cards. If you were not correct, the other player takes both cards.

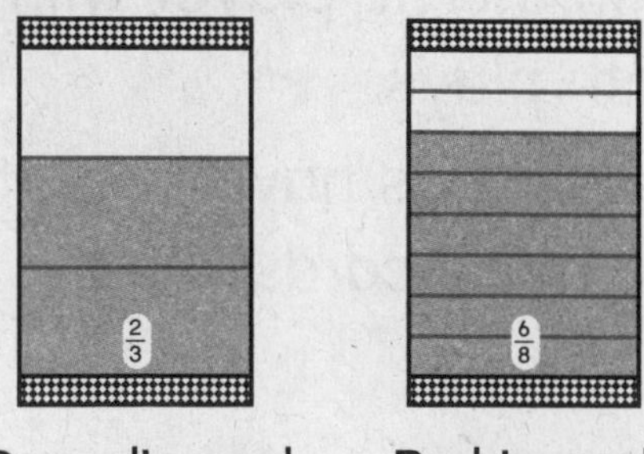

Denzel's card Barb's card

Less of Denzel's card is shaded: $\frac{2}{3}$ is less than $\frac{6}{8}$. Denzel takes both cards.

Date Time

LESSON 8•6

Math Boxes

1. Fill in the missing numbers.

992	
	1,003

2. There are

_____ minutes in an hour.

_____ hours in a day.

_____ days in a week.

_____ weeks in a year.

3. 368

The value of 3 is _____.

The value of 6 is _____.

The value of 8 is _____.

MRB 10

4. 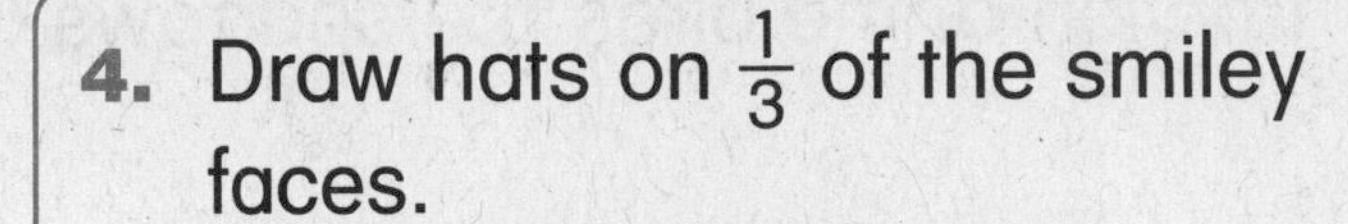
Draw hats on $\frac{1}{3}$ of the smiley faces.

5. Fill in the circle next to the best answer. A school bus is about:

Ⓐ 180 cm long.

Ⓑ 18 m long.

Ⓒ 18 in long.

Ⓓ 180 m long.

6. □ = 1 sq cm

Area = _____ sq cm

Date Time

Fraction Number Stories

Solve these number stories. To help, you can use pennies or other counters, or you can draw pictures.

1. Mark has 4 shirts to wear. 3 of them have short sleeves. What fraction of the shirts have short sleeves?

   ____

2. 8 birds are sitting on a tree branch. 6 of the birds are sparrows. What fraction of the birds are sparrows?

   ____

3. June has 15 fish in her fish tank. $\frac{1}{3}$ of the fish are guppies. How many guppies does she have?

   ____

Try This

4. Sam ate $\frac{0}{5}$ of a candy bar. How much of the candy bar did he eat?

   ____

5. If you were thirsty, would you rather have $\frac{2}{2}$ of a carton of milk or $\frac{4}{4}$ of that same carton? Explain.

   ____

   ____

   ____

   ____

Date Time

LESSON 8·7

Math Boxes

1. **Baskets Made by 2nd Graders**

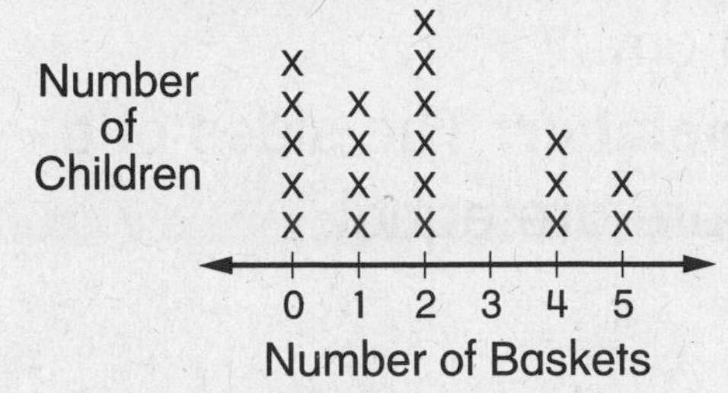

What was the most frequent number of baskets made (the *mode*)? ______

2. What fraction of dots is circled? Circle the best answer.

A $\frac{1}{2}$

B $\frac{1}{4}$

C $\frac{1}{3}$

D $\frac{2}{4}$

3. Fill in the missing numbers.

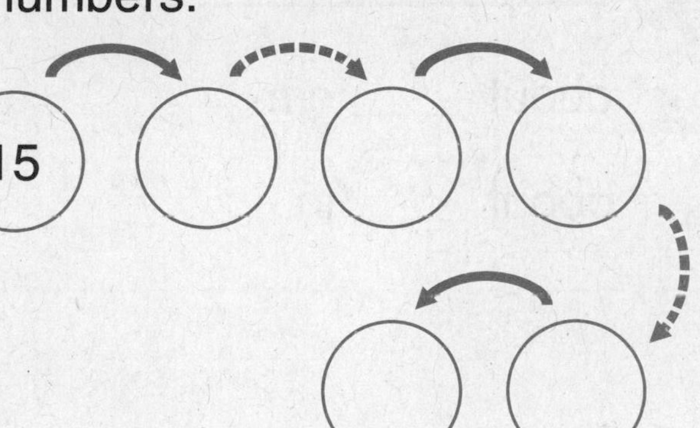

4. Complete the diagram. Then write a number model.

Quantity
82

Quantity
39

______
Difference

____

5. Draw a line segment that is 6 cm long. Divide the line segment into 3 equal parts.

Each part = ______ cm

6. Measure each side to the nearest cm. Find the perimeter.

The perimeter is ______ cm.

MRB
68

Date Time

LESSON 8•8

Math Boxes

1. Circle. How much does a can of soda hold?

   1 ounce

   1 gallon

   36 liters

   12 ounces

2. Draw a square with a perimeter of 8 cm.

   Remember: The sides of a square are equal.

3. Draw a rectangle. Measure each side to the nearest cm.

   about _____ cm

   about _____ cm

   about _____ cm

   about _____ cm

4. Measure the length of this line.

   ____

   about _____ cm

   about _____ in.

5. Match the items to the weights.

1 cat	1 ounce
3 envelopes	1 pound
1 book	7 pounds

6. □ = 1 sq cm

   Area = _____ sq cm

Date Time

LESSON 9·1 Meters

Materials ☐ meterstick

Directions

Record each step in the table below.

1. Choose a distance.
2. Estimate the distance in meters.
3. Use a meterstick to measure the distance to the nearest meter. Compare this measurement to your estimate.

Distance I Estimated and Measured	My Estimate	My Meterstick Measurement
	about ______ meters	about ______ meters
	about ______ meters	about ______ meters
	about ______ meters	about ______ meters
	about ______ meters	about ______ meters
	about ______ meters	about ______ meters

Date Time

LESSON 9•1 Comparing Measurements

Work with a partner. Measure your height, head size, and shoe length to the nearest centimeter. For each measurement, choose a tool to use. You may use a ruler, meterstick, or tape measure.

1. Height

I am about ______ centimeters tall.

My partner is about ______ centimeters tall.

Who is taller? ____________________

How much taller? ______ cm

2. Head size (the distance around your head)

My head is about ______ centimeters.

My partner's head size is about ______ centimeters.

Who has the larger head size? ____________________

How much larger? ______ cm

3. Shoe length

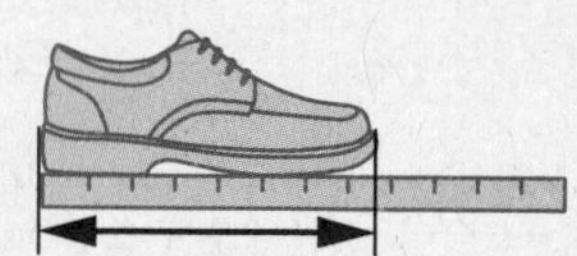

My shoe is about ______ centimeters.

My partner's shoe is about ______ centimeters.

Who has the longer shoe length? ____________________

How much longer? ______ cm

LESSON 9·1

Math Boxes

1. Make a square array with 25 pennies. How many pennies are in each row?

____ pennies

2. Write $<$, $>$, or $=$.

$4 + 5 + 6$ ______ $3 + 5 + 7$

$7 + 5 + 9$ ______ $6 + 6 + 8$

$2 + 11 + 4$ ______ $7 + 1 + 9$

$15 + 7 + 5$ ______ $9 + 9 + 9$

3. Circle the parallel lines.

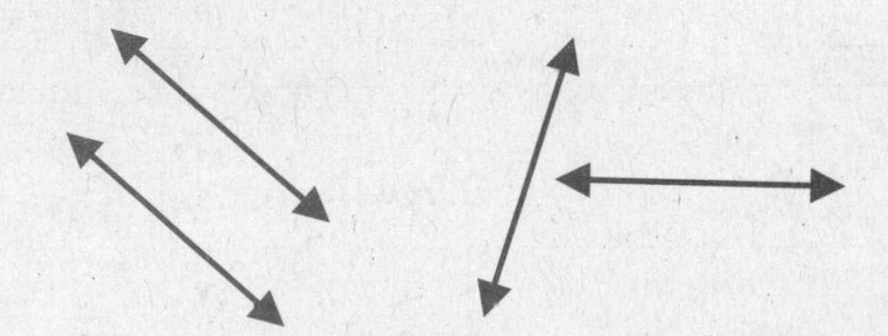

4.

246 228 273
209 298

Unit
yards

The median number of yards is ______.

5. Use your Pattern-Block Template. Trace a shape and draw 1 line of symmetry.

6. Count by quarters to \$3.00.

\$0.50, ________, ________,

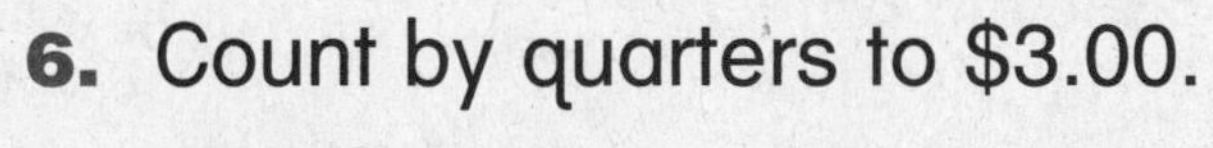

________, ________, ________,

________, ________, ________,

________, ________

Date Time

LESSON 9•2

Units of Linear Measure

Materials ☐ foot and decimeter rulers from *Math Masters,* p. 255

Directions

1. Select two objects or distances.
2. Estimate the length of each object or distance in feet, inches, decimeters, and centimeters. Write your estimates.
3. Measure the length of each object or distance in feet, inches, decimeters, and centimeters. Write your measurements.

First Object or Distance	My Estimates	Actual Measurements
	about _____ ft	about _____ ft
	about _____ in.	about _____ in.
	about _____ dm	about _____ dm
	about _____ cm	about _____ cm

Second Object or Distance	My Estimates	Actual Measurements
	about _____ ft	about _____ ft
	about _____ in.	about _____ in.
	about _____ dm	about _____ dm
	about _____ cm	about _____ cm

LESSON 9·2

Math Boxes

1. Fill in the missing numbers.

			1,250
		1,259	
	1,268		

2. Draw a line segment that is 8 cm long. Now draw a line segment 5 cm shorter.

3. Dillon leaped 32 inches. Marcus leaped 27 inches. How many more inches did Dillon leap? ______ inches

Fill in the diagram.

Quantity

Quantity

______
Difference

4. What is the chance that you will have two birthdays this year? Choose the best answer.

- ⬭ unlikely
- ⬭ likely
- ⬭ certain
- ⬭ impossible

5. 3 insects. 6 legs per insect. How many legs in all?

______ legs

Fill in the diagram and write a number model.

insects	legs per insect	legs in all

____

6. Solve.

______ pennies = \$2.00

______ nickels = \$2.00

______ dimes = \$2.00

______ quarters = \$2.00

Date Time

LESSON 9·3

Measuring Lengths with a Ruler

Work with a partner. Use your ruler to measure the length of each object to the nearest inch and centimeter.

1. large paper clip 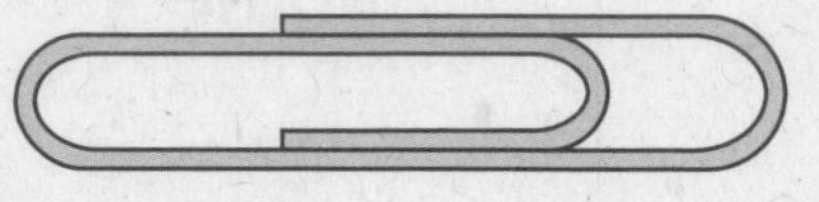

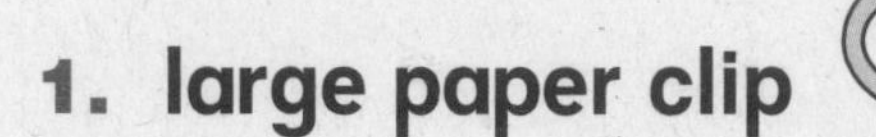

about ______ inches long about ______ centimeters long

2. pencil 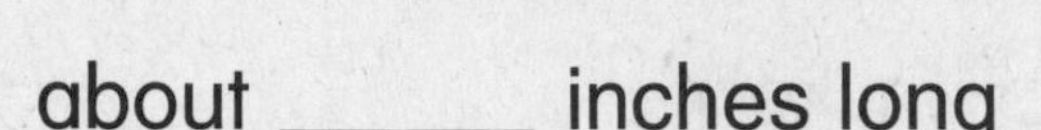

about ______ inches long about ______ centimeters long

3. nail

about ______ inch long about ______ centimeters long

4. How much longer is the pencil than the large paper clip?

______ inches ______ centimeters

Try This

Measure to the nearest $\frac{1}{2}$-inch and $\frac{1}{2}$-centimeter.

5. small paper clip 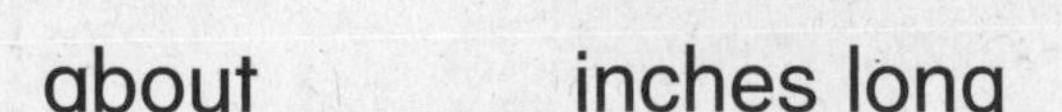

about ______ inches long about ______ centimeters long

Date Time

Math Boxes

1. Make a square array with 36 pennies. How many pennies in each row?

____ pennies

2. Complete each number model.

Unit

______ > 199

372 > ______

______ < 2,424

5,269 < ______

3. Draw a quadrangle. Make 2 sides parallel.

MRB 51 55

4. 264 246 310 277 301

Unit
meters

Find the median number of meters. Circle the best answer.

A. 264 **B.** 277

C. 301 **D.** 310

5. Find 2 lines of symmetry.

6. How much in all?

______

(D) (D)
(Q) (Q) (Q) (N)

[$20] [$5]
[$10] [$5]

Date Time

LESSON 9•4

Distance Around and Perimeter

Measure the distance around the following to the nearest centimeter.

1. Your neck: _____ cm

2. Your ankle: _____ cm

Measure the distance around two other objects to the nearest centimeter.

3. Object: ____________________ Measurement: _____ cm

4. Object: ____________________ Measurement: _____ cm

Measure each side of the figure to the nearest inch. Write the length next to each side. Then find the perimeter.

5.

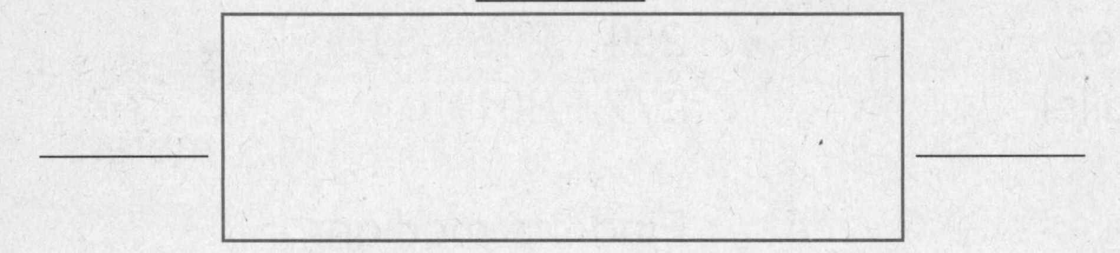

Perimeter: _____ inches

Try This

6.

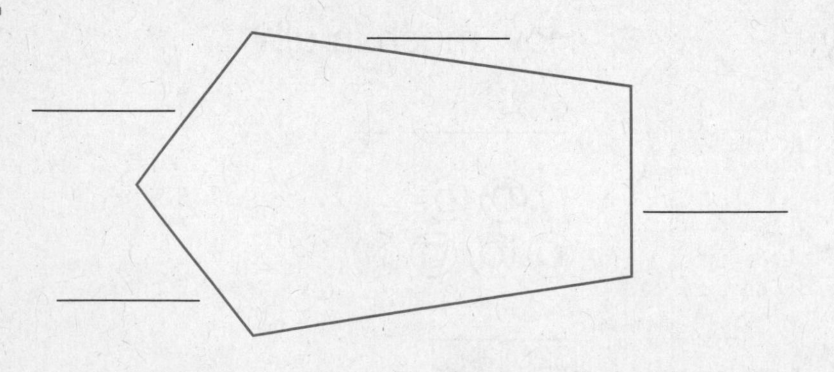

Perimeter: _____ inches

Date Time

LESSON 9·4

Math Boxes

1. Fill in the missing numbers.

	1,789	
	1,799	

2. Draw a line segment that is $3\frac{1}{2}$ inches long.

Now draw a line segment that is 1 inch shorter.

3. The Jays scored 63 points. The Gulls scored 46 points. How many more points did the Jays score? _____ points

Fill in the diagram.

Quantity

Quantity

_____
Difference

4. What is the chance that you will fly in a spaceship today? Circle your answer.

impossible

certain

likely

unlikely

5. 9 cars. Each has 4 tires. How many tires in all?

cars	tires per car	tires in all

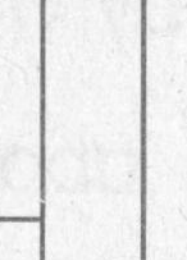

6. Use Ⓟ, Ⓝ, Ⓓ, and Ⓠ. Show $1.79.

Date Time

LESSON 9•5

Driving in the West

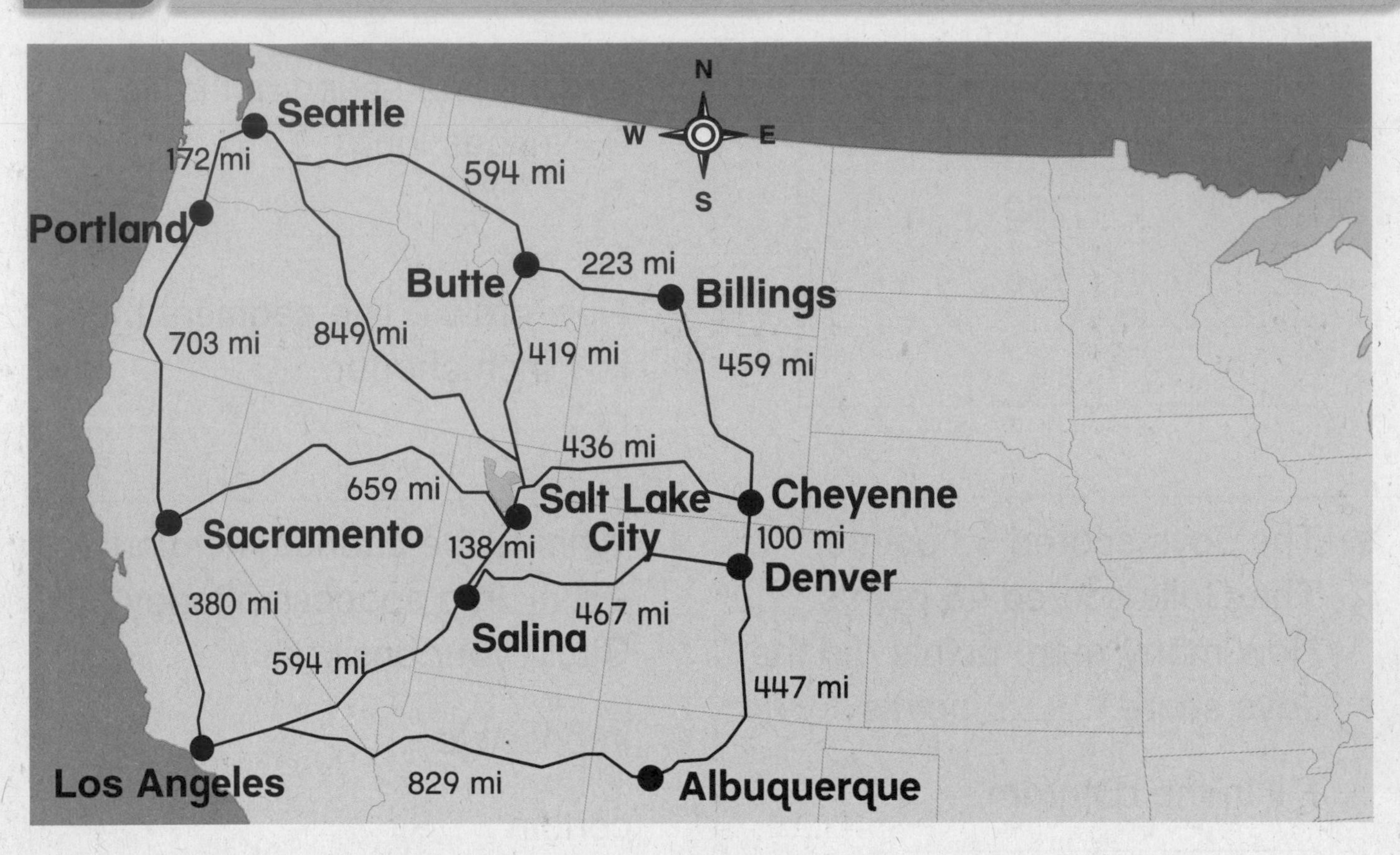

1. What is the shortest route from Seattle to Albuquerque?

____

____

2. Put a check mark in front of the longer trip.

   ______ Salt Lake City to Billings by way of Butte

   ______ Salt Lake City to Billings by way of Cheyenne

   How much longer is that trip? about ______ miles longer

Date ______________________ Time ______________________

LESSON 9·5

Addition and Subtraction Practice

Make a ballpark estimate. Solve. Compare your answer to your estimate.

1. Ballpark estimate:

____

$45 + 68 =$ _______

2. Ballpark estimate:

____

$143 + 78 =$ _______

3. Ballpark estimate:

____

$158 + 233 =$ _______

4. Ballpark estimate:

____

$74 - 49 =$ _______

5. Ballpark estimate:

____

$133 - 86 =$ _______

6. Ballpark estimate:

____

$256 - 147 =$ _______

Date Time

LESSON 9·5

Math Boxes

1. Write 5 names for $\frac{1}{2}$. Use your Fraction Cards if you need help.

$\frac{1}{2}$

2. Write even or odd.

126 ________

311 ________

109 ________

430 ________

3. 17 pieces of gum are shared equally. Each child gets 3 pieces.

How many children are sharing?

______ children

How many pieces are left over?

______ pieces

4. Rosita had $0.39 and found $0.57 more. How much does she have now? Estimate your answer and then use partial sums to solve.

Estimate:

_______ + _______ = _______

Answer: _______

5. In Pensacola, Florida, the temperature is 82°F. In Portland, Maine, the temperature is 64°F. What is the difference? Fill in the circle next to the best answer.

Ⓐ 22°F
Ⓑ 17°F
Ⓒ 20°F
Ⓓ 18°F

6. A number has:

7 thousands
8 tens
5 ten-thousands
1 one
0 hundreds

Write the number. __________

Date Time

LESSON 9·6

Math Message

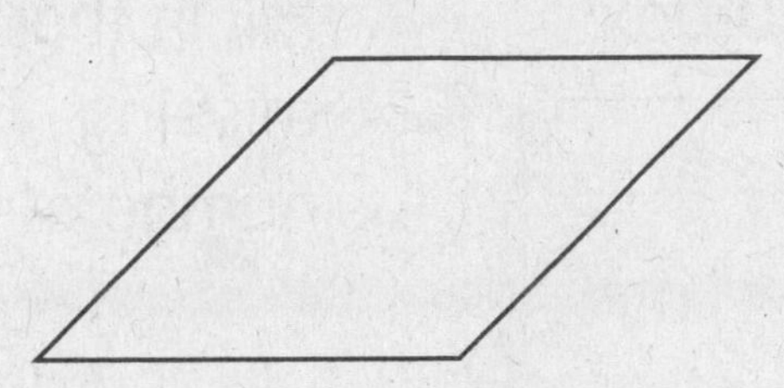

Estimate: Which shape is the "biggest" (has the largest area)? Circle it.

Think: How might you measure the shapes to find out?

Exploration A: Which Cylinder Holds More?

Which holds more macaroni—the tall and narrow cylinder or the short and wide cylinder?

My prediction: ______________________________

Actual result: ______________________________

Exploration B: Measuring Area

The area of my tracing of the deck of cards is about _____ square centimeters.

The area of my tracing of the deck of cards is about _____ square inches.

I traced ______________________________.

It has an area of about ______________________________.
(unit)

Date Time

LESSON 9•6

Math Boxes

1. Write 5 names in the 90-box.

MRB 16

2. Fill in the missing numbers.

+2 −5

98

MRB 98 99

3. Solve. Show your work.

$$\begin{array}{r} 27 \\ +\ 56 \\ \hline \end{array}$$

4.

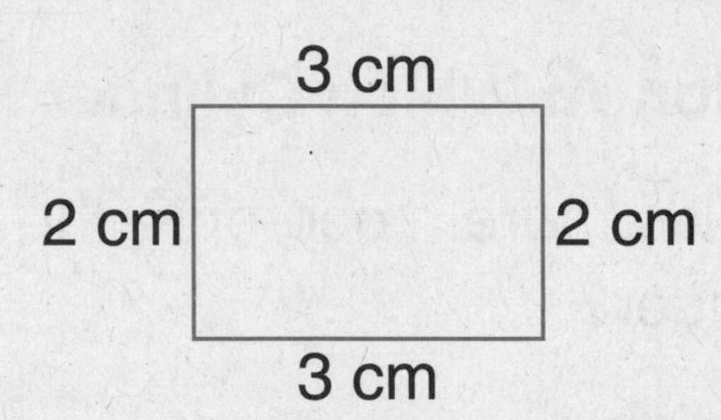

Perimeter = ______ cm

5. Write a number model for a ballpark estimate. Then solve.

Ballpark estimate:

____

$$\begin{array}{r} 68 \\ +\ 34 \\ \hline \end{array}$$

6. The total cost is 60¢. You pay with a $1 bill.

How much change do you get?

______

Show the change using Ⓠ, Ⓓ, and Ⓝ.

Date Time

LESSON 9•7

Finding Area

For each shape, count the square centimeters to find the area.

1. 

Area = ______ sq cm

2. 

Area = ______ sq cm

Use the tick marks to draw lines to show square units. Then count the squares to find the area.

3. 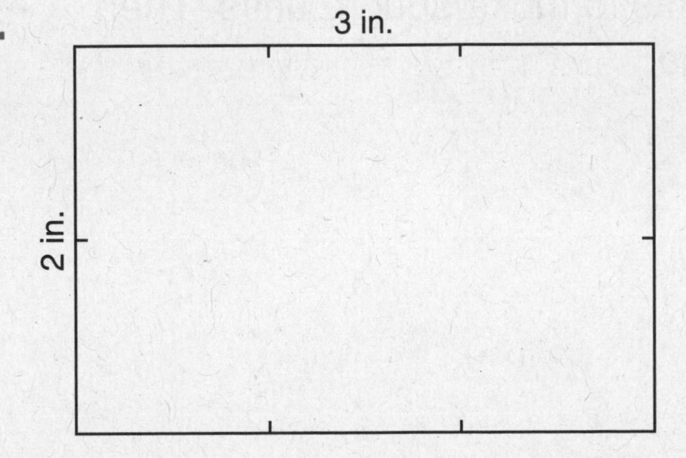

Area = ______ sq in.

4. 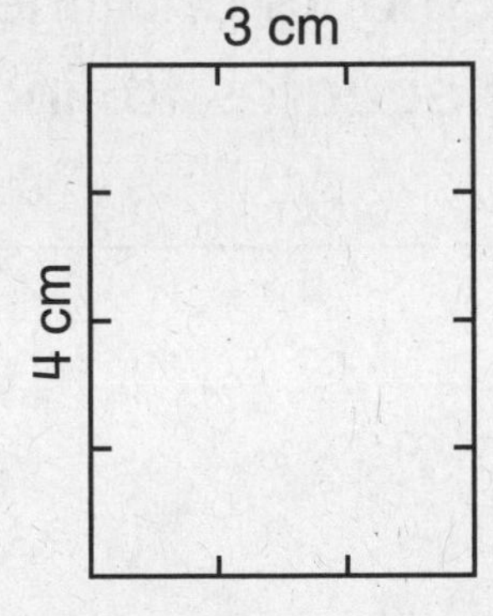

Area = ______ sq cm

Date Time

LESSON 9•7

Finding Area *continued*

For each shape, use the tick marks to draw lines to show square units. Then count the squares to find the area.

5.

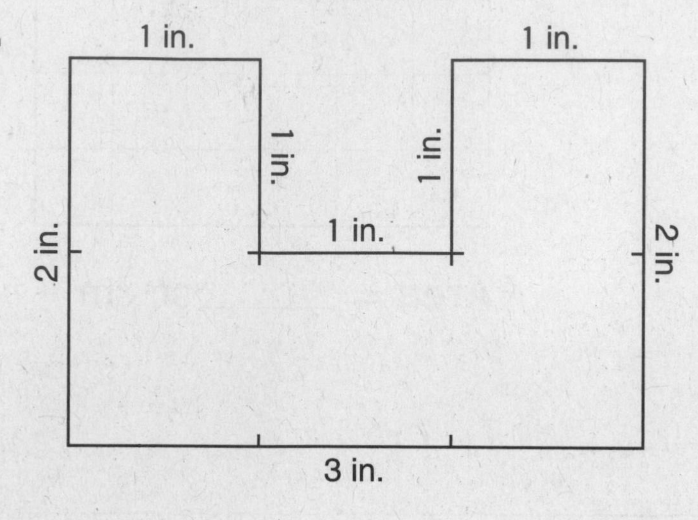

Area = ______ sq in.

6.

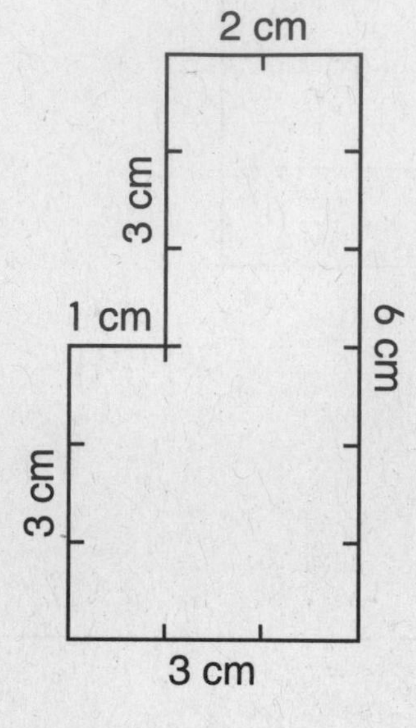

Area = ______ sq cm

Try This

7. Draw tick marks and line segments to make square units. Then count the squares to find the area.

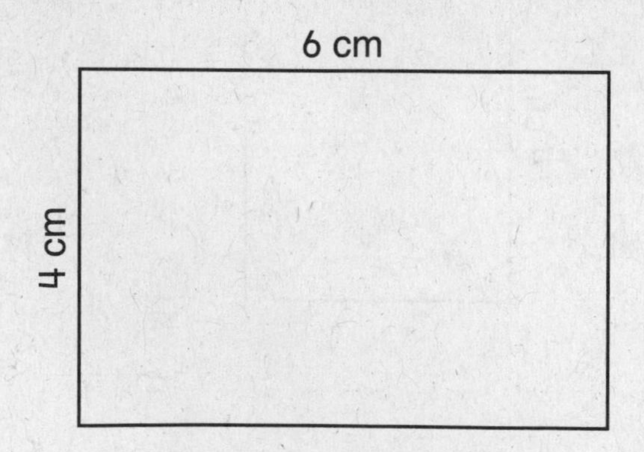

Area = ______ sq cm

Date Time

LESSON 9·7

Math Boxes

1. Draw two ways to show $\frac{2}{3}$.

2. Write 3 even numbers larger than 100.

_______, _______, _______

Write 3 odd numbers smaller than 100.

_______, _______, _______

3. Get 36 counters. Share them equally among 4 children.

How many counters does each child get? _____ counters

How many are left over?

_____ counters

4. I have $2.00. Can I buy 4 bags of chips for $0.55 each?

_____

5. Solve.

Unit

$386 - 40 =$ _____

_____ $= 198 - 60$

$259 - 40 =$ _____

_____ $= 243 - 20$

6. In 43,692, the value of

4 is __________.

3 is __________.

6 is __________.

9 is __________.

2 is __________.

Date _______ Time _______

LESSON 9•8

Equivalent Units of Capacity

Complete.

U.S. Customary Units of Capacity

_____ pint = 1 cup

1 pint = _____ cups

_____ pints = 1 quart

_____ quarts = 1 half-gallon

_____ half-gallons = 1 gallon

Metric Units of Capacity

1 liter = _____ milliliters

$\frac{1}{2}$ liter = _____ milliliters

1. How many quarts are in 1 gallon? _____ quarts

2. How many cups are in 1 quart? _____ cups

In a half-gallon? _____ cups In 1 gallon? _____ cups

3. How many pints are in a half-gallon? _____ pints

In a gallon? _____ pints

"What's My Rule?"

4.

Rule
1 qt = 2 pt

qt	pt
2	
3	
	10
8	

5.

Rule
1 gal = 8 pt

gal	pt
2	
3	
	40
	80

Date Time

LESSON 9•8

Math Boxes

1. Write 5 names for 130.

130

2. Fill in the missing numbers.

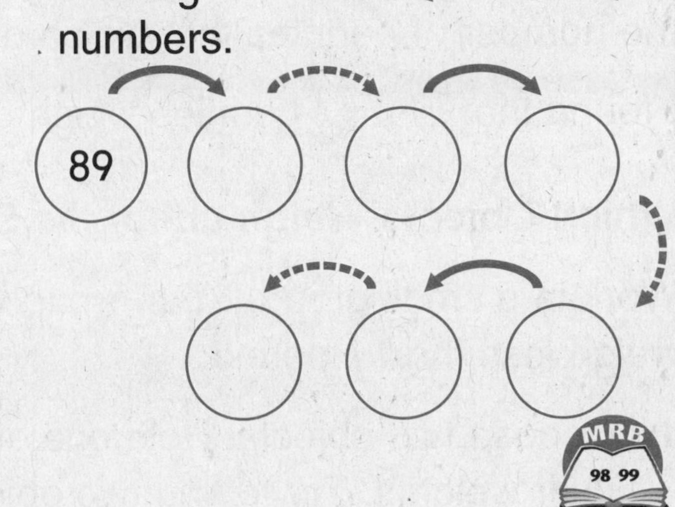

3. Solve. Show your work.

$$\begin{array}{r} 49 \\ +\ 23 \\ \hline \end{array}$$

4.

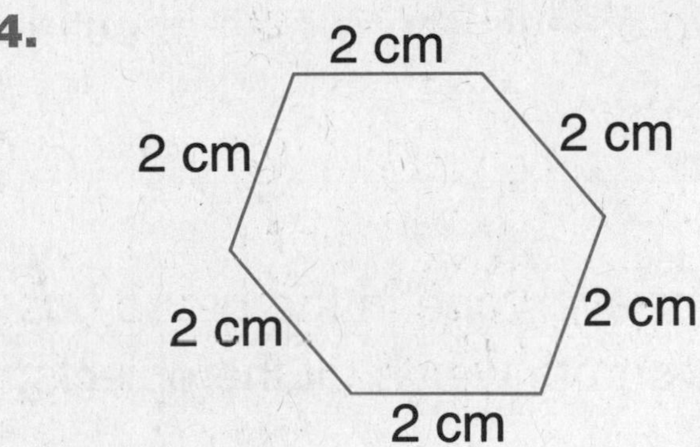

Perimeter = ______ cm

5. Estimate. Then solve.

Estimate:

______ + ______ = ______

$$\begin{array}{r} 57 \\ +\ 48 \\ \hline \end{array}$$

6. The total cost is $1.50. You pay with a $5 bill. How much change do you get?

Fill in the circle next to the best answer.

Ⓐ $6.50 Ⓑ $2.50

Ⓒ $4.00 Ⓓ $3.50

Date _______________ Time _______________

Weight

Weighing Pennies

Use a spring scale, letter scale, or diet scale to weigh pennies. Find the number of pennies that weigh about 1 ounce.

I found that ______ pennies weigh about 1 ounce.

Which Objects Weigh about the Same?

Work in a small group. Your group will be given several objects that weigh less than 1 pound.

1. Choose two objects. Hold one object in each hand and compare their weights. Try to find two objects that weigh about the same.

   Two objects that weigh about the same:

   ______________________ ______________________

2. After everyone in the group has chosen two objects that weigh about the same, weigh all the objects. Record the weights below.

Object	**Weight** (include unit)
______________________	______________
______________________	______________
______________________	______________
______________________	______________
______________________	______________
______________________	______________

Date Time

LESSON 9•9

Math Boxes

1. Use your Fraction Cards. Find another name for $\frac{3}{4}$. Circle the best answer.

   A $\frac{1}{2}$ B $\frac{3}{8}$

   C $\frac{6}{8}$ D $\frac{2}{3}$

2. Write 2 even 4-digit numbers.

   ________ ________

   Write 2 odd 4-digit numbers.

   ________ ________

3. Use counters to solve.

   18 orange slices are shared equally. Each child gets 4 slices.

   How many children are sharing?

   ______ children

   How many slices are left?

   ______ slices

4. Fill in the missing amount.

   I had 38¢.

   I spent ______¢.

   I have 15¢ left.

5. Solve.

Unit

   22 − 14 = ______

   62 − 14 = ______

   162 − 14 = ______

   ______ = 292 − 14

   ______ = 402 − 14

6. In 96,527, the value of

   5 is ____________.

   6 is ____________.

   7 is ____________.

   2 is ____________.

   9 is ____________.

LESSON 9•10

Math Boxes

1. I have 85¢. How many 5¢ pieces of candy can I buy?

______ pieces of candy

2. 67,248

The value of 8 is ________.

The value of 2 is ________.

The value of 7 is ________.

The value of 6 is ________.

The value of 4 is ________.

3. How much money?

$10 $5
Ⓠ Ⓝ Ⓝ Ⓟ

$________

4. The total cost of Neena's lunch is $7.50.

She paid with a $10 bill.

How much change will she get?

$ ________

5. Justin had $0.92 and found $0.21 more. How much does he have now? Estimate your answer and then solve.

Estimate:

________ + ________ = ________

Answer: ________

6. Use Ⓟ, Ⓝ, Ⓓ, and Ⓠ. Show $2.20.

Date Time

LESSON 10•1

Math Boxes

1. Double.

25¢ ________

55¢ ________

65¢ ________

85¢ ________

2. 15 children. $\frac{1}{3}$ are boys.

How many are boys? ________

How many are girls? ________

3. Count by 1000s.

________; 2,728; ________;

________; ________; ________;

________; ________; ________

4. Solve.

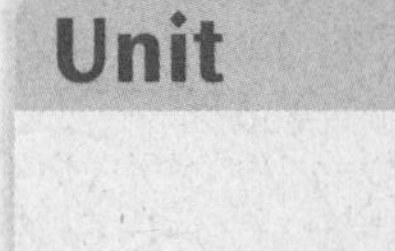

$5 + 3 =$ ________

$50 + 30 =$ ________

$$\begin{array}{r} 6 \\ +\ 3 \\ \hline \end{array} \qquad \begin{array}{r} 60 \\ +\ 30 \\ \hline \end{array}$$

5. Draw a rhombus. Make each side 2 cm long.

6. How many dots are in this 5-by-5 array?

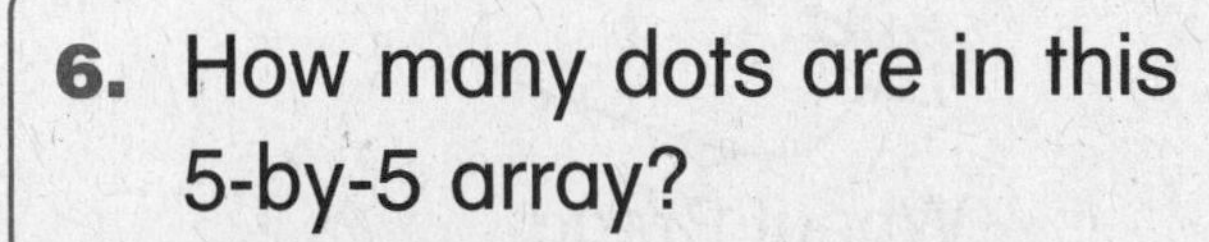

________ dots in all

Date Time

Good Buys Poster

Fruit/Vegetables Group

Seedless Grapes
99¢ lb

Carrots
1-lb bag
3/$1.00

Plums
69¢ lb

Oranges
$1.49 lb

Bananas
59¢ lb

Watermelons
$2.99 ea.

Celery
59¢ lb

Grain Group

Wheat Bread
16 oz
99¢

Saltines
1 lb
69¢

Hamburger Buns
16 oz
69¢

Meat Group

Pork & Beans
16 oz
2/89¢

Peanut Butter
18-oz jar
$1.29

Ground Beef
$1.99 lb

Chunk Light
Tuna
6.5 oz

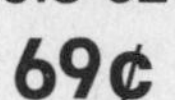

69¢

Lunch Meat
1-lb package
$1.39

Milk Group

Gallon
Milk
$2.39

American
Cheese
8 oz
$1.49

6-pack
Yogurt
$2.09

Miscellaneous Items

Mayonnaise
32 oz
$1.99

Catsup
32 oz
$1.09

Grape Jelly
2-lb jar
$1.69

Date Time

Ways to Pay

Complete Problem 1. For Problems 2 and 3 choose two items from the Good Buys Poster on page 230. List the items and how much they cost in the table below.

For each item:

- Count out coins and bills to show several different ways of paying for each item.
- Record two ways by drawing coins and bills in the table. Use Ⓠ, Ⓓ, Ⓝ, Ⓟ, and $1.

Example:

You buy 1 pound of bananas. They cost 59¢ a pound. You pay with:
Ⓠ Ⓠ Ⓝ Ⓟ Ⓟ Ⓟ Ⓟ or Ⓓ Ⓓ Ⓓ Ⓓ Ⓓ Ⓝ Ⓟ Ⓟ Ⓟ Ⓟ

1. You buy *oranges*. Cost: $1.49 Pay with | or	**2.** You buy ________. Cost: ________ Pay with | or
3. You buy ________. Cost: ________ Pay with | or	**Try This** **4.** You buy ________ and ________. Cost: ________ Pay with | or

LESSON 10•2

Word Values

Pretend the letters of the alphabet have the dollar values shown in the table. For example, the letter **g** is worth \$7; the letter **v** is worth \$22. The word **jet** is worth \$10 + \$5 + \$20 = \$35.

	a	b	c	d	e	f	g	h	i	j	k	l	m
Value	\$1	\$2	\$3	\$4	\$5	\$6	\$7	\$8	\$9	\$10	\$11	\$12	\$13
	n	o	p	q	r	s	t	u	v	w	x	y	z
Value	\$14	\$15	\$16	\$17	\$18	\$19	\$20	\$21	\$22	\$23	\$24	\$25	\$26

1. Which is worth more, **dog** or **cat**? ____________

2. Which is worth more, **whale** or **zebra?** ____________

3. How much is your first name worth? ________________

4. Write 2 spelling words you are trying to learn. Find their values.

   Word: ________________ Value: \$________________

   Word: ________________ Value: \$________________

5. What is the cheapest word you can make? It must have at least 2 letters.

   Word: ________________ Value: \$________________

6. What is the most expensive word you can make?

   Word: ________________ Value: \$________________

Try This

7. Think of the letter values as dimes. For example, **m** is worth 13 dimes; **b** is worth 2 dimes. Find out how much each word is worth.

   dog: \$______ cat: \$______ zebra: \$______ whale: \$______

   candy: \$______ your last name: \$________________

Date Time

LESSON 10•2

Math Boxes

1. Write <, >, or =.

1,257 ______ 2,157

7,925 ______ 5,297

10,129 ______ 1,129

2. Circle the answer.

$2.88 is closer to:
$2.80 or $2.90

$5.61 is closer to:
$5.60 or $5.70

$1.97 is closer to:
$1.90 or $2.00

3. Put the heights in order. Find the median height.

Unit
inches

48 44 37 54 39

____, ____, ____, ____, ____

The median height is

______ inches.

4. What is the temperature? Circle the best answer.

A. 85° F

B. 86° F

C. 83° F

D. 76° F

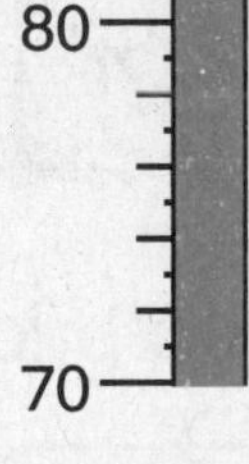

5. Draw the hour and minute hands to show the time 20 minutes later than 6:15.

What time does the clock show now?

____:____

6. You have 21 pennies to share equally among 3 children. How many pennies does each child get?

______ pennies

How many are left over?

______ pennies

Date ______________________ Time ______________________

LESSON 10•3

Calculator Dollars and Cents

To enter $4.27 into your calculator, press (4) (.) (2) (7).

To enter 35¢ into your calculator, press (.) (3) (5).

1. Enter $3.58 into your calculator. The display shows ____________.

2. Enter the following amounts into your calculator.

   Record what the display shows.
   Don't forget to clear between each entry.

Price	Display
$2.75	____________
$1.69	____________
$12.32	____________

Make up prices that are more than $1.00.

3. Enter 68¢ into your calculator. The display shows ____________.

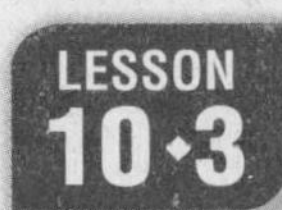

LESSON 10·3 Calculator Dollars and Cents *continued*

4. Enter the following amounts into your calculator.
Record what you see in the display.

Price	Display
$0.10	________
$0.26	________
$0.09	________

Make up prices that are less than $1.00.

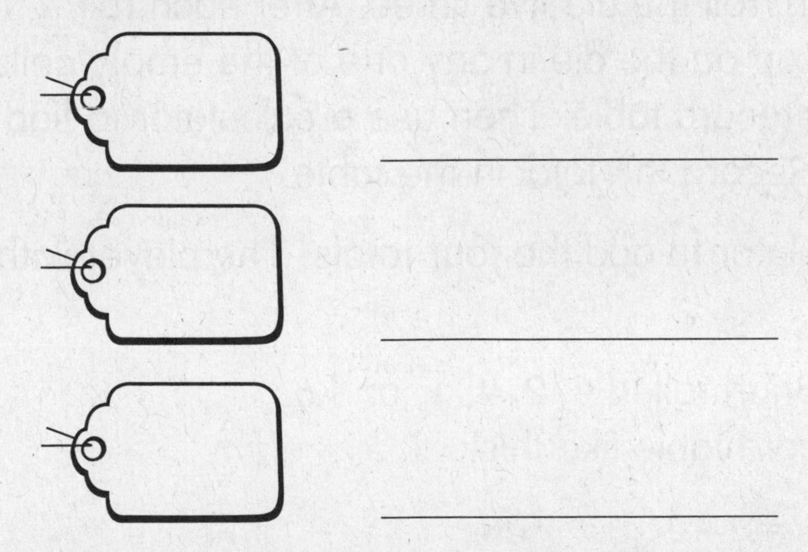

5. Use your calculator to add $1.55 and $0.25.

What does the display show? ______

Explain what happened. ______________________________

Date Time

LESSON 10·3

Pick-a-Coin Directions

Materials ☐ 1 die ☐ calculator for each player

☐ *Pick-a-Coin* record table for each player (*Math Journal 2,* p. 237 or *Math Masters*, p. 469)

Players 2 to 4

Skill Add dollar bill and coin combinations

Object of the Game To add the largest value

Summary

Players roll a die. The numbers that come up are used as numbers of coins and dollar bills. Players try to make collections of coins and bills with the largest value.

Directions

Take turns. When it is your turn, roll the die five times. After each roll, record the number that comes up on the die in any one of the empty cells in the row for that turn on your record table. Then use a calculator to find the total amount for that turn. Record the total in the table.

After four turns, use your calculator to add the four totals. The player with the largest Grand Total wins.

Example: On his first turn, Brian rolled 4, 2, 4, 1, and 6. He filled in his record table like this:

***Pick-a-Coin* Record Table**

	Ⓟ	Ⓝ	Ⓓ	Ⓠ	$1	Total
1st turn	2	1	4	4	6	$7.47
2nd turn						$___.___
3rd turn						$___.___
4th turn						$___.___
				Grand Total		$___.___

Date Time

LESSON 10•3

Pick-a-Coin Record Tables

	(P)	(N)	(D)	(Q)	[$1]	Total
1st turn						$___ . _____
2nd turn						$___ . _____
3rd turn						$___ . _____
4th turn						$___ . _____
					Grand Total	$___ . _____

	(P)	(N)	(D)	(Q)	[$1]	Total
1st turn						$___ . _____
2nd turn						$___ . _____
3rd turn						$___ . _____
4th turn						$___ . _____
					Grand Total	$___ . _____

	(P)	(N)	(D)	(Q)	[$1]	Total
1st turn						$___ . _____
2nd turn						$___ . _____
3rd turn						$___ . _____
4th turn						$___ . _____
					Grand Total	$___ . _____

Date Time

LESSON 10•3

Finding Area

Find the area of the shapes below. For Problem 1, count the square centimeters to find the area. For Problems 2–4, draw lines to show the square units. Use the tick marks as a guide. Then count the squares to find the area.

1.

Area = ________ sq cm

2.

2 in.

2 in.

Area = ________ sq in.

3.

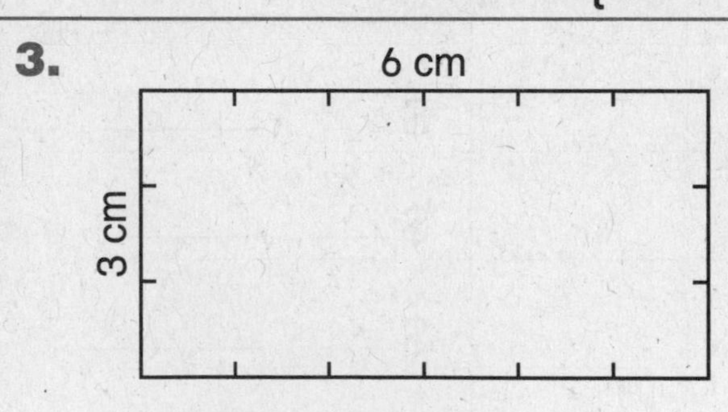

Area = ________ sq cm

4.

5 cm

2 cm

4 cm

3 cm

2 cm

2 cm

Area = ________ sq cm

Try This

5. Draw tick marks and line segments to make square units. Then count the squares to find the area.

5 cm

3 cm

Area = ________ sq cm

Date Time

LESSON 10·3

Math Boxes

1. ________ pennies = $3.00

   ________ nickels = $3.00

   ________ dimes = $3.00

   ________ quarters = $3.00

2. Count 20 pennies.

   $\frac{1}{2}$ = ________ pennies

   $\frac{1}{4}$ = ________ pennies

   $\frac{1}{5}$ = ________ pennies

3. Complete the frames.

4. Solve.

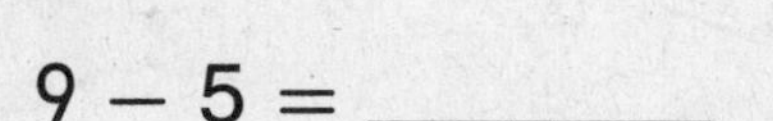

Unit

9 − 5 = ________

________ = 90 − 50

900 − 500 = ________

________ = 9,000 − 5,000

5. Match.

5 ft	3 yd
24 in.	60 in.
9 ft	2 ft

MRB 67

6. Draw an 8-by-4 array.

   How many in all? ________

LESSON 10•4

Then-and-Now Poster

1897

Cheddar Cheese
$\frac{1}{2}$ lb
6¢

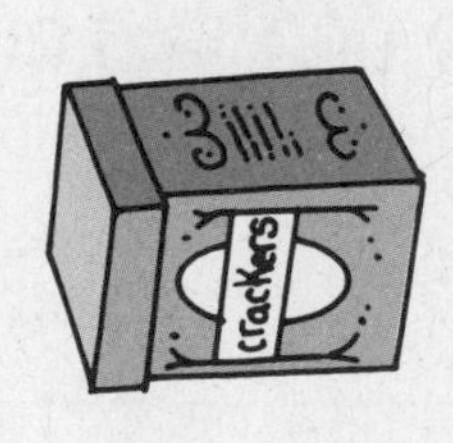

Crackers
1 lb
6¢

Raisins
1 lb
10¢

Grape Jelly
2 lb
28¢

Catsup
32 oz/1 qt
25¢

20-Inch Girl's Bicycle
$29.00

Child's Wagon
Large Size –15" × 30"
$1.65

Harmonica
Ten Double Holes
45¢

Now

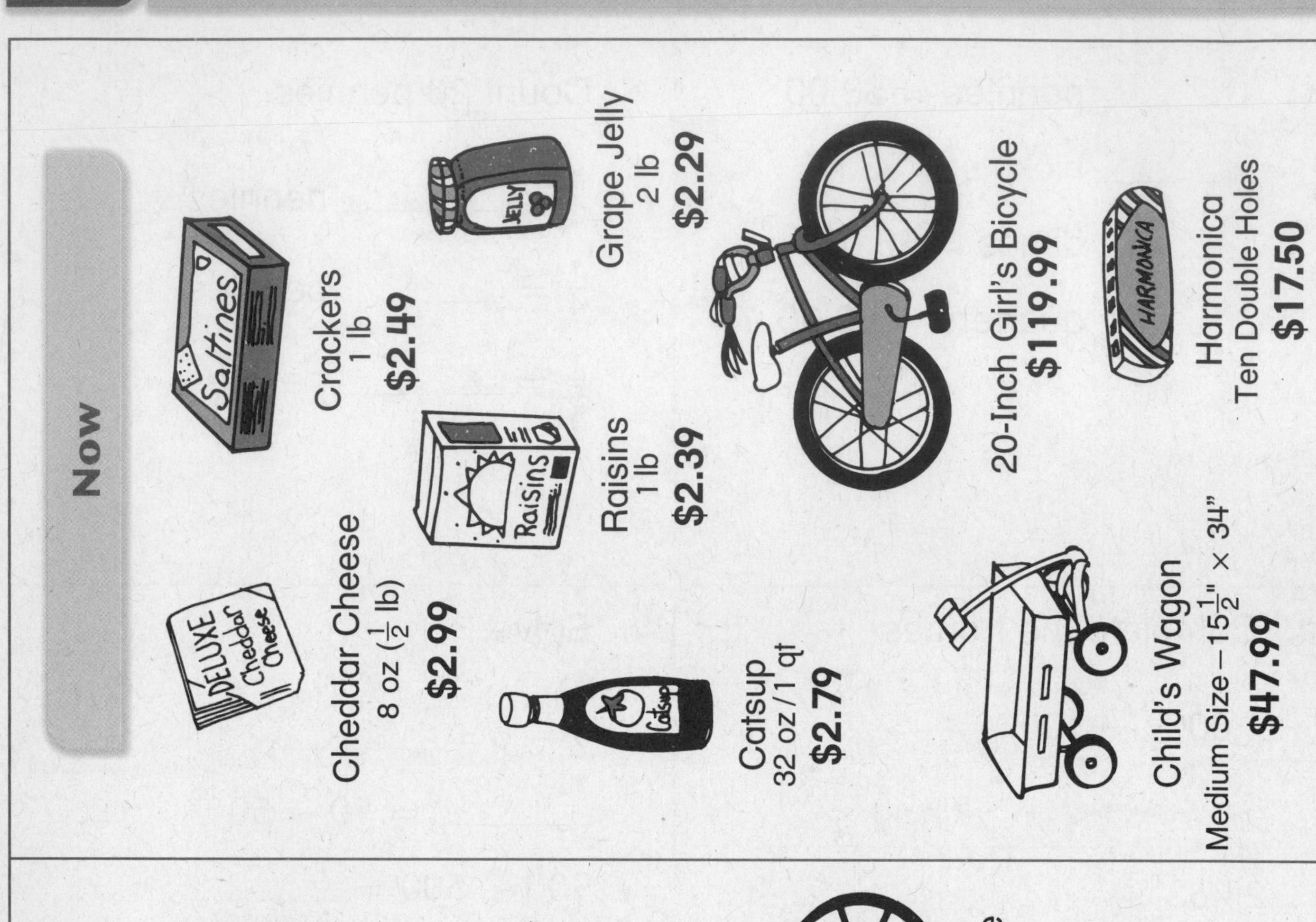

Cheddar Cheese
8 oz ($\frac{1}{2}$ lb)
$2.99

Crackers
1 lb
$2.49

Raisins
1 lb
$2.39

Grape Jelly
2 lb
$2.29

Catsup
32 oz/1 qt
$2.79

20-Inch Girl's Bicycle
$119.99

Child's Wagon
Medium Size –$15\frac{1}{2}$" × 34"
$47.99

Harmonica
Ten Double Holes
$17.50

LESSON 10·4

Then-and-Now Prices

Use your calculator.

1. How much did a 20-inch bicycle cost in 1897? ____________

   How much does it cost now? ____________

   How much more does it cost now? ____________

2. How much more does a pound of cheese cost now

   than it did in 1897? ____________

3. In 1897, raisins were packed in cartons. Each carton contained 24 one-pound boxes. How much did a

   24-pound carton cost then? ____________

   How much would it cost now? ____________

4. Which item had the biggest price increase from then to now?

   ____________________________ had the biggest price increase.

   How much more does it cost now? ____________

5. Our Own Problems about Then-and-Now:

   ____

   ____

   ____

   ____

   ____

   ____

Date Time

LESSON 10•4

Math Boxes

1. Write <, >, or =.

1,292 + 10 ____ 1,285 + 15

3,791 + 7 ____ 3,799 + 7

5,020 + 100 ____ 5,125 + 25

2. Fill in the blanks to estimate the total cost.

\$2.43 + \$0.39 is about

____ + ____ = ____

\$0.88 + \$0.67 is about

____ + ____ = ____

3. Arrange the numbers in order.

Find the median.

98 56 143 172 81

____, ____, ____,

____, ____

The median is ____.

4. Show 55°F.

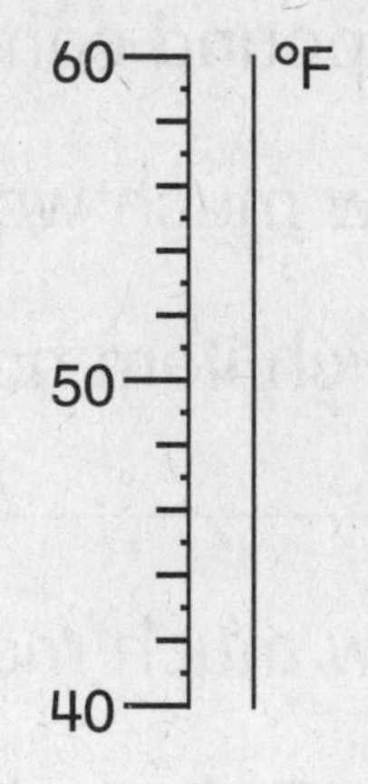

5. It is 6:15. Draw the hour and minute hands to show the time 15 minutes later.

What time does the clock show?

___:___

6. Use counters to solve.

35 blocks are shared equally among 3 children. How many blocks does each child get?

____ blocks

How many blocks are left over?

____ blocks

Date Time

LESSON 10•5

Estimating and Buying Food

Choose items to buy from the Good Buys Poster on journal page 230. For each purchase:

- Record the items on the sales slip.
- Write the price of each item on the sales slip.
- Estimate the total cost and record it.
- Your partner then uses a calculator to find the exact total cost and writes it on the sales slip.

Purchase 1 **Price**

Items:

________________ $ ___.___

________________ $ ___.___

Estimated cost: about $ ___.___

Exact total cost: $ ___.___

Purchase 2 **Price**

Items:

________________ $ ___.___

________________ $ ___.___

Estimated cost: about $ ___.___

Exact total cost: $ ___.___

Purchase 3 **Price**

Items:

________________ $ ___.___

________________ $ ___.___

Estimated cost: about $ ___.___

Exact total cost: $ ___.___

Date Time

LESSON 10•5

Math Boxes

1. Show $1.73 in two different ways. Use Ⓟ, Ⓝ, Ⓓ, and Ⓠ.

____

____

____

____

2. A notebook costs $2.00. A pen costs 50¢ less than the notebook. How much do they cost together?

Answer: ________

3. Fill in the rule and the missing numbers.

Rule

in	out
1,342	2,342
3,019	4,019
4,650	
	6,700

4. Solve.

Unit
km

$6 + 5 =$ ________

$60 + 50 =$ ________

$600 + 500 =$ ________

$6,000 + 5,000 =$ ________

5. Write <, >, or =.

1 qt ____ 1 pt

3 c ____ 1 gal

1 qt ____ 4 c

1 gal ____ 5 pt

6. There are 3 drink boxes per pack. How many packs are needed to serve 25 second graders and 2 teachers one drink box each? Draw an array. Circle the best answer.

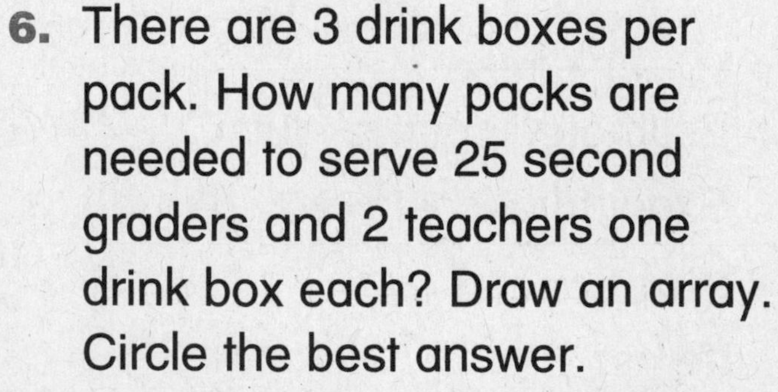

______ packs are needed.

A. 30 **B.** 8 **C.** 9 **D.** 10

Date Time

LESSON 10·6

Math Boxes

1. What number is shown by the blocks?

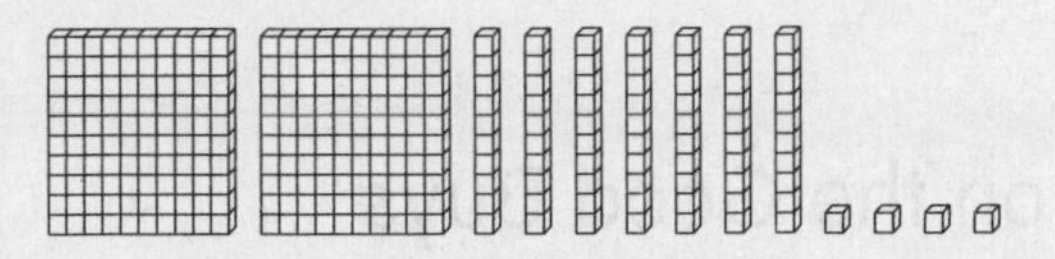

____

2. Joe has $1.00 and spends 65¢. How much change will he get?

____

3. What is the temperature?

_______ °F

Is it warm or cold outside?

____

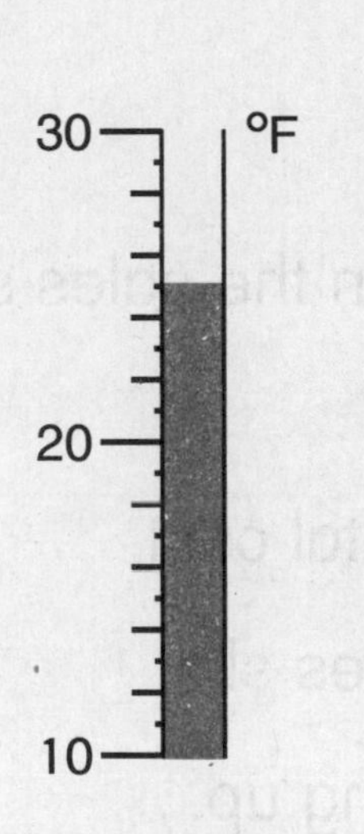

4. You buy some stickers for $1.89. Show 2 ways to pay. Use Ⓟ, Ⓝ, Ⓓ, Ⓠ, and [$1].

5. Write the names of 3 objects that are shaped like cylinders.

____

____

____

6. How many stars in all?

★★★★★
★★★★★
★★★★★

_______ stars

Fill in the multiplication diagram.

rows	stars per row	stars in all

MRB 112–113

LESSON 10·6

Making Change

Work with a partner. Use your tool-kit coins and bills. One of you is the shopper. The other is the clerk.

The shopper does the following:

- Chooses one item from each food group on the Good Buys Poster on journal page 230.
- Lists these items on the Good Buys sales slip in the shopper's journal on page 247.
- Writes the cost of each item on the sales slip.
- Estimates the total cost of all the items and writes it on the sales slip.
- Pays with a $10 bill.
- Estimates the change and writes it on the sales slip.

The clerk does the following:

- Uses a calculator to find the exact total cost.
- Writes the exact total cost on the sales slip.
- Gives the shopper change by counting up.
- Writes the exact change from $10.00 on the sales slip.

Change roles and repeat.

Date Time

LESSON 10·6

Making Change *continued*

The Good Buys Store Sales Slip		
	Item	**Cost**
Fruit/vegetables group	______	$ ___ . ___
Grain group	______	$ ___ . ___
Meat group	______	$ ___ . ___
Milk group	______	$ ___ . ___
Miscellaneous items	______	$ ___ . ___
Estimated total cost		$ ___ . ___
Estimated change from $10.00		$ ___ . ___
Exact total cost		$ ___ . ___
Exact change from $10.00		$ ___ . ___

Date Time

LESSON 10·7

The Area of My Handprint

The area of each ☐ is 1 square centimeter. Other ways to write *square centimeter* are sq cm and cm^2.

The area of my handprint is _____ square centimeters, or _____ sq cm.

Date Time

LESSON 10·7

The Area of My Footprint

The area of each ☐ is 1 square centimeter. Other ways to write *square centimeter* are sq cm and cm^2.

The area of my footprint is _____ square centimeters, or _____ sq cm.

LESSON 10·7 Worktables

Use a Pattern-Block Template to draw each of the different-size and different-shape worktables you made. Write the name of each shape next to your drawing.

Date Time

LESSON 10·7

Geoboard Dot Paper

1.

2.

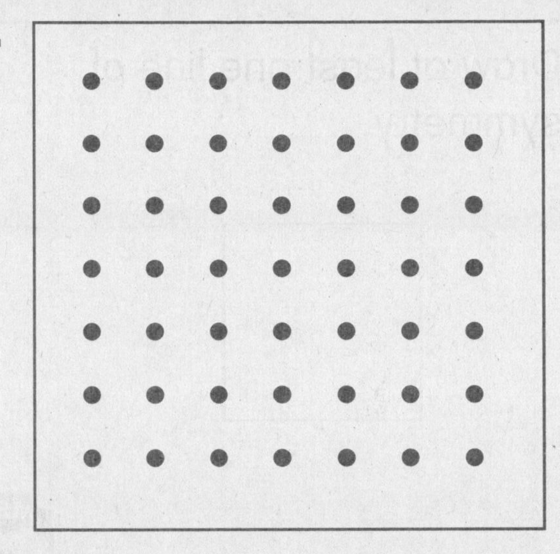

3.

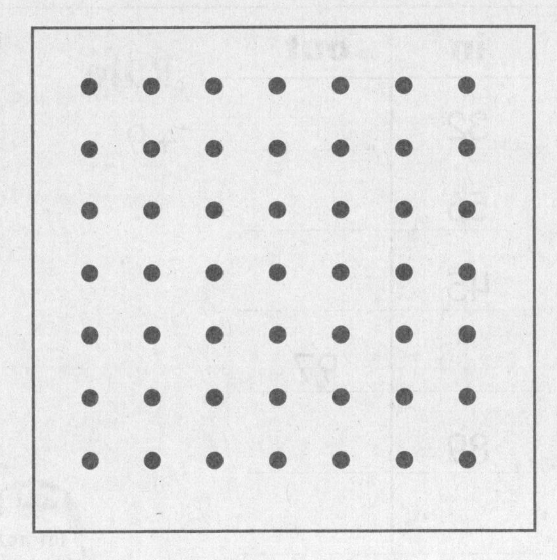

4.

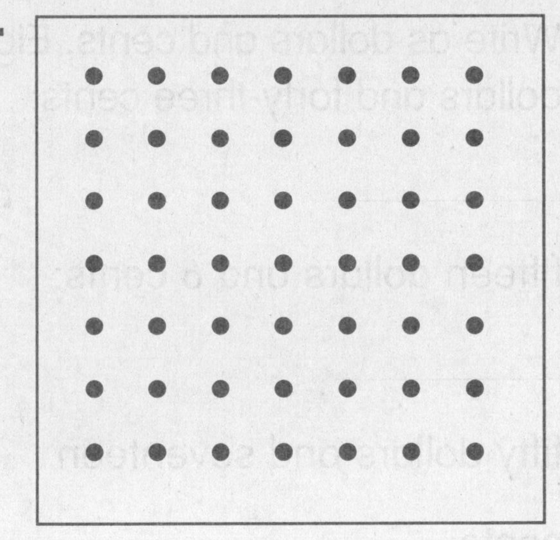

5.

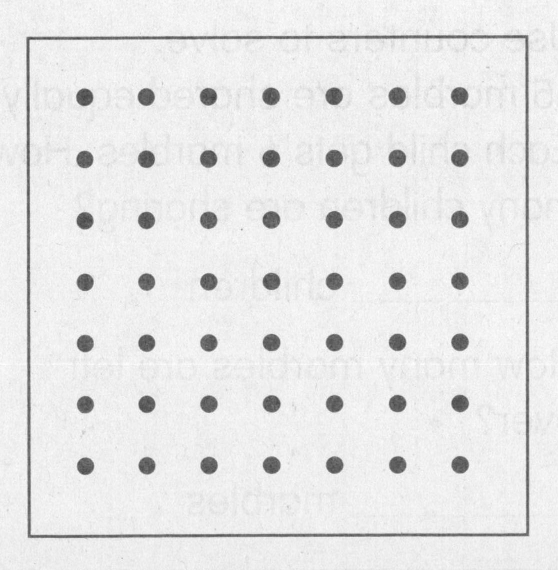

6.

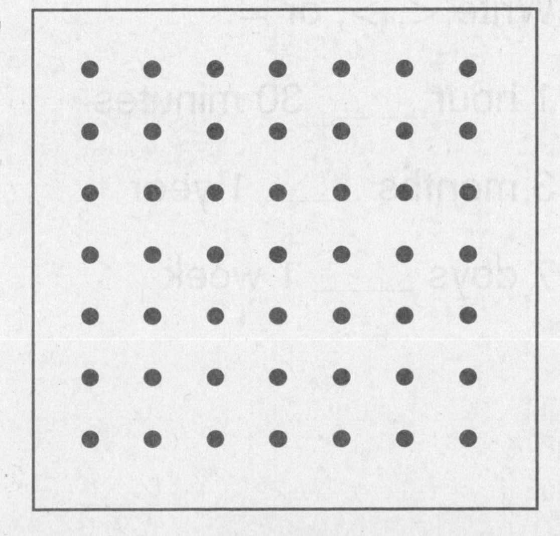

Date Time

LESSON 10·7

Math Boxes

1. Draw at least one line of symmetry.

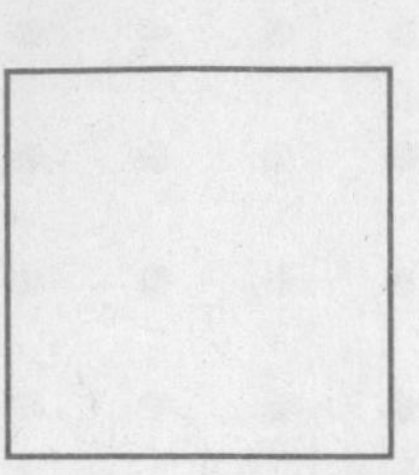

2. A pencil costs 95¢. An eraser costs 35¢ more than the pencil. How much do the pencil and the eraser cost together?

Answer: ________

3. Write as dollars and cents. Eight dollars and forty-three cents:

____

fifteen dollars and 6 cents:

____

fifty dollars and seventeen cents: ________

4.

in	out
32	
56	
45	
	97
89	

Rule
+9

5. Write <, >, or =.

1 hour ____ 30 minutes

3 months ____ 1 year

7 days ____ 1 week

6. Use counters to solve.
15 marbles are shared equally. Each child gets 6 marbles. How many children are sharing?

________ children

How many marbles are left over?

________ marbles

LESSON 10·8

Money Exchange Game Directions

Materials ☐ 1 six-sided die

☐ 1 ten- or twelve-sided die

☐ 24 pennies, 39 dimes, thirty-nine $1 bills, and one $10 bill per player

Players 2 or 3

Skill Make exchanges between coins and bills

Object of the Game To be the first to trade for $10

Directions

1. Each player puts 12 pennies, 12 dimes, twelve $1 bills, and one $10 bill in the bank.
2. Players take turns. Players use a six-sided die to represent pennies. Players use a ten- or twelve-sided die to represent dimes.
3. Each player
   - rolls the dice.
   - takes from the bank the number of pennies and dimes shown on the faces of the dice.
   - puts the coins in the correct columns on his or her Place-Value Mat on journal page 254.
4. Whenever possible, a player replaces 10 coins or bills of a lower denomination with a coin or bill of the next higher denomination.
5. The first player to trade for a $10 bill wins.

If there is a time limit, the winner is the player with the largest number on the mat when time is up.

Date Time

LESSON 10•8

Place-Value Mat

(P) pennies — 1s	
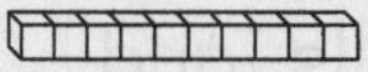 (D) dimes — 10s	
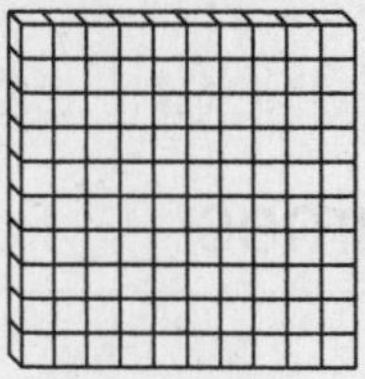 $1 dollars — 100s	
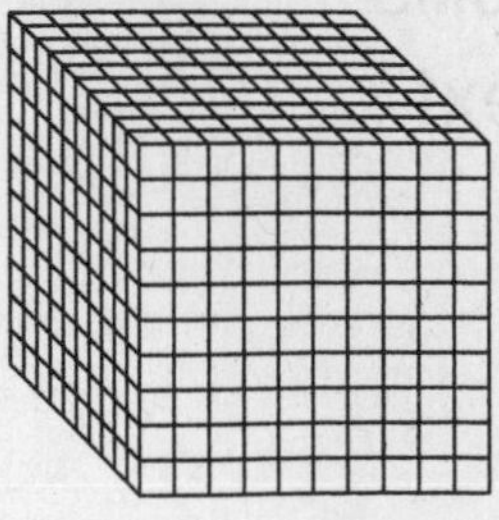 $10 — 1,000s	

Date   Time

LESSON 10·8

Ballpark Estimates

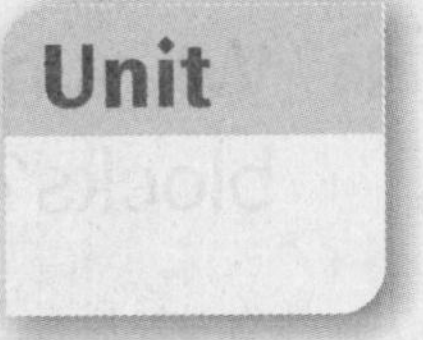

Fill in the unit box. Then, for each problem:

Make a ballpark estimate before you add.

Write a number model for your estimate.

Use your calculator and solve the problem. Write the exact answer in the box.

Compare your estimate to your answer.

1. Ballpark estimate: ______ 148 + 27	**2.** Ballpark estimate: ______ 163 + 32	**3.** Ballpark estimate: ______ 133 + 35
4. Ballpark estimate: ______ 143 + 41	**5.** Ballpark estimate: ______ 184 + 23	**6.** Ballpark estimate: ______ 154 + 183

Date Time

LESSON 10·8

Math Boxes

1. What number is shown by the blocks?

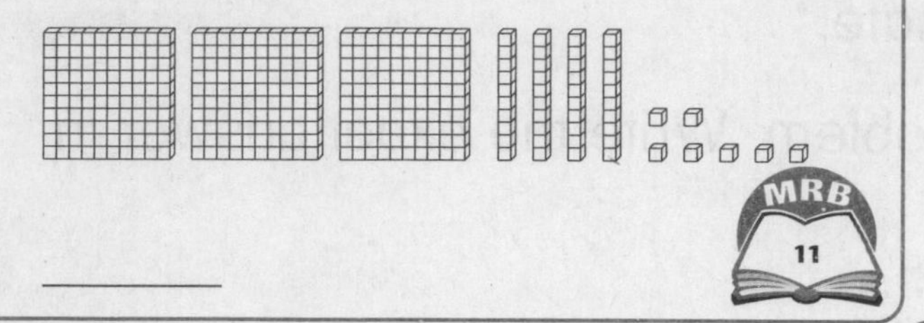

______

MRB 11

2. Kelly has $10. Her lunch total was $8.75. How much change will she get?

3. In the morning, it was 62°F. By afternoon, the temperature was 75°F. How much did the temperature rise? ______

Start | Change | End

______

Number model: ______

MRB 116–118

4. Cross out the names that don't belong.

10¢

ten cents, $\frac{1}{10}$ of a dollar,

$10.00, Ⓓ, ⓃⓃ, $0.01,

$\frac{1}{100}$ of a dollar,

$\frac{1}{2}$ of a dollar

MRB 88–90

5. Which object is shaped like a cone? Circle the best answer.

A shoe box

B party hat

C paper towel roll

D globe

MRB 57

6. 4 ladybugs. 5 spots on each ladybug. How many spots?

______

Fill in the diagram and write a number model.

lady bugs	spots per lady bug	spots in all

______

MRB 112 113

Date Time

LESSON 10•9

Math Boxes

1. Use your Pattern-Block Template.

Trace a trapezoid. Draw the line of symmetry.

2. Color $\frac{3}{4}$ of the circle.

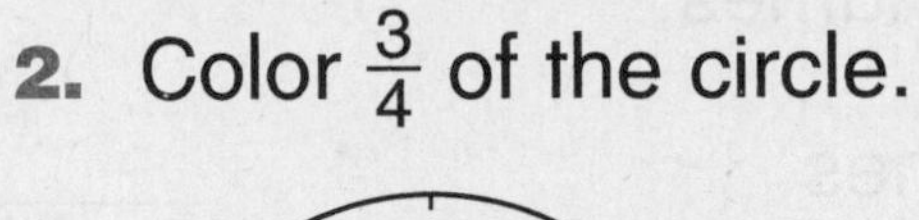

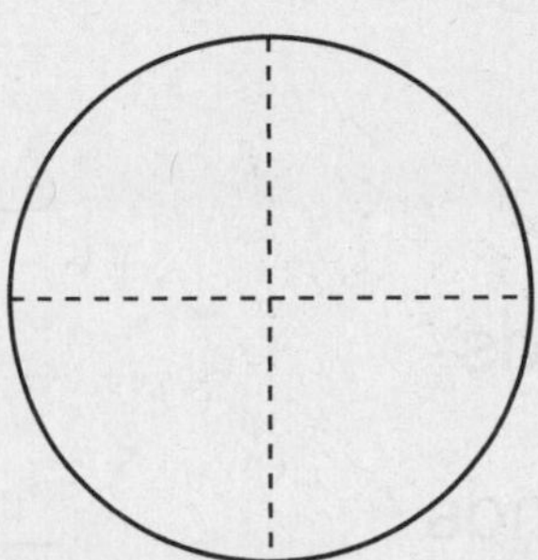

What fraction of the circle is not colored? ______

3. Write the amounts.

Five thousand six hundred eight dollars and twelve cents

____

Two hundred sixteen dollars and sixty-eight cents

____

Three hundred nine dollars and five cents

____

4.

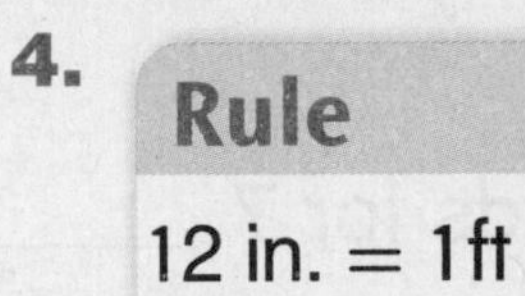

in	out
6	
	2
48	

5. ______ hours in a day

______ days in a week

______ months in a year

______ weeks in a year

6. Use counters to solve. 26 children. 2 children for each computer. How many computers?

______ computers

Date Time

LESSON 10·10

Place Value

1. Match names.

A. 5 ones ________ 50

B. 5 tens ________ 500

C. 5 hundreds ________ 50,000

D. 5 thousands ________ 5

E. 5 ten-thousands ________ 5,000

Fill in the blanks. Write ones, tens, hundreds, thousands, or ten-thousands.

2. The 7 in 187 stands for 7 ________________.

3. The 2 in 2,785 stands for 2 ________________.

4. The 3 in 4,239 stands for 3 ________________.

5. The 0 in 13,409 stands for 0 ________________.

6. The 5 in 58,047 stands for 5 ________________.

Continue.

7. 364; 365; 366; ________; ________; ________

8. 996; 997; 998; ________; ________; ________

9. 1,796; 1,797; 1,798; ________; ________; ________

10. 1,996; 1,997; 1,998; ________; ________; ________

11. 9,996; 9,997; 9,998; ________; ________; ________

Date Time

LESSON 10·10

Math Boxes

1. Circle the digit in the 1,000s place.

4, 6 9 4

2 9, 4 0 0

2 0, 0 0 4

5, 0 1 9

Read each number to a partner.

2. I have a 5-dollar bill. I spend \$4.38. How much change do I get?

______

3. Show 25° C on the thermometer.

Is it good weather to go ice skating or to go to the beach?

______

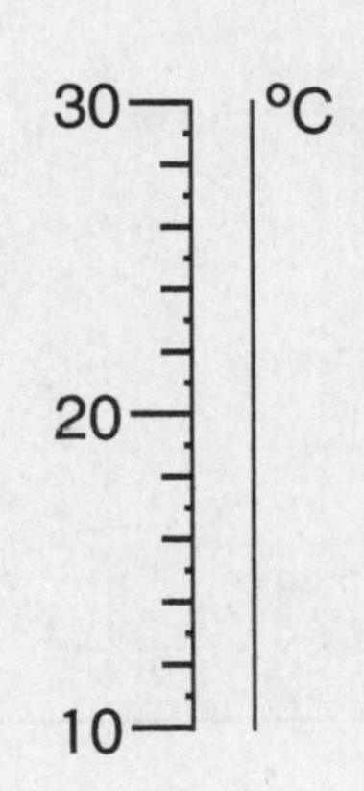

4. Write 5 names for \$0.75.

______

______

______

______

______

5. Write the names of 3 objects shaped like rectangular prisms.

______

______

______

6. 5 wagons. 4 wheels on each wagon. How many wheels?

______ wheels

wagons	wheels per wagon	wheels in all

MRB 112 113

Date Time

LESSON 10•11

Parentheses Puzzles

Parentheses can make a big difference in a problem.

Example:

$15 - 5 + 3 = ?$

$(15 - 5) + 3 = (10) + 3 = 13$; but

$15 - (5 + 3) = 15 - (8) = 7$

Solve problems containing parentheses.

1. $7 + (8 - 3) =$ _____

2. $(4 + 11) - 6 =$ _____

3. $8 + (13 - 9) =$ _____

4. _____ $= (12 + 8) - 16$

5. $140 - (20 + 80) =$ _____

6. _____ $= (30 + 40) - 70$

Put in parentheses to solve the puzzles.

7. $12 - 4 + 6 = 14$

8. $15 - 9 - 4 = 10$

9. $140 - 60 + 30 = 110$

10. $500 = 400 - 100 + 200$

11. $3 \times 2 + 5 = 11$

12. $2 \times 5 - 5 = 0$

Date Time

LESSON 10·11

Math Boxes

1. Choose any shape on your Pattern-Block Template that has at least 2 lines of symmetry. Trace the shape and draw the lines of symmetry.

2. Circle $\frac{3}{12}$

What fraction of dots is not circled? Circle the best answer.

A $\frac{9}{3}$ **B** $\frac{9}{12}$

C $\frac{6}{12}$ **D** $\frac{3}{9}$

Write another name for this fraction ________.

3. Use your calculator to find the total.

$1 $1 $1 $1 = $________

Ⓠ Ⓠ Ⓠ = $________

Ⓓ Ⓓ Ⓓ Ⓓ Ⓓ = $________

Ⓝ Ⓝ Ⓝ Ⓝ Ⓝ Ⓝ Ⓝ = $________

Total = $________

4.

Rule
5Ⓝ = 1Ⓠ

Ⓝ	Ⓠ
	3
	4
30	
	10
	20

5. ________ months = 1 year

________ months = 2 years

________ months = 3 years

________ months = 4 years

6. You have 18 pieces of gum to share equally. If each child gets 4 pieces, how many children are sharing?

________ children

How many pieces of gum are left over?

________ pieces of gum

LESSON 10•12

Math Boxes

1. 6 lily pads. 3 frogs per lily pad. How many frogs in all?

______ frogs

Fill in the diagram and write a number model.

lily pads	frogs per lily pad	frogs in all

____

2. Draw a 7-by-3 array.

How many in all? ______

3. 17 magazines are shared equally among 5 children. Draw a picture to help you.

Each child gets ___ magazines.

There are ___ magazines left.

4. 8 books per shelf. 4 shelves. Fill in the diagram and solve.

shelves	books per shelf	books in all

There are ______ books.

5. This is a ______ -by- ______ array.

How many dots in all?

______ dots

6. 11 balloons are shared equally among 4 children. How many balloons does each child get?

______ balloons

How many are left over?

______ balloons

Date Time

LESSON 11•1 Math Boxes

1. Circle the unit that makes sense.

Grandma's house is

5 ______ away. km dm

Mona's goldfish is

8 ______ long. cm m

Ahmed's dad is

68 ______ tall. cm in.

2. Circle the fraction that is bigger. Use your Fraction Cards to help.

$\frac{2}{3}$ or $\frac{2}{2}$ $\frac{4}{5}$ or $\frac{2}{5}$

$\frac{2}{8}$ or $\frac{5}{6}$ $\frac{3}{6}$ or $\frac{1}{4}$

3. Write a number with 5 in the thousands place.

____

What is the value of the digit 5 in your number?

____

MRB 10 11

4. Draw a 3-by-6 array.

How many in all? ______

5. Had a $10 bill.
Spent $8.90.
How much change?

____

6. Write the fact family.

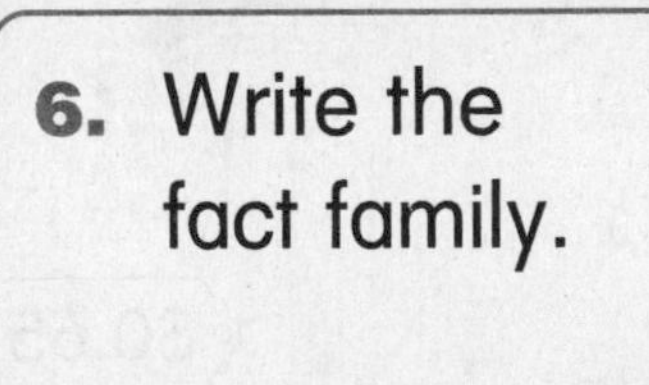

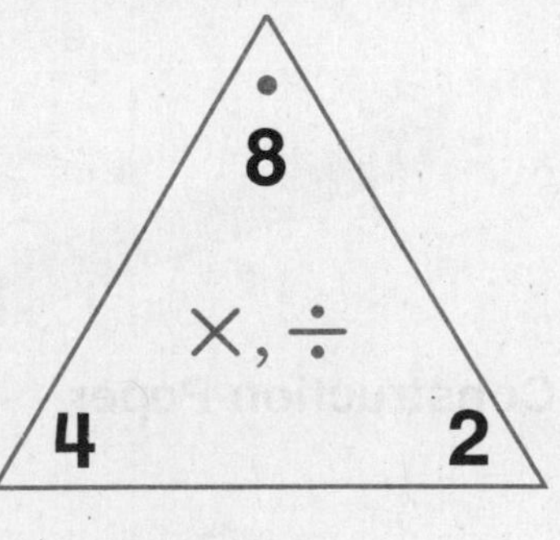

______ × ______ = ______

______ × ______ = ______

______ ÷ ______ = ______

______ ÷ ______ = ______

MRB 38

Art Supply Poster

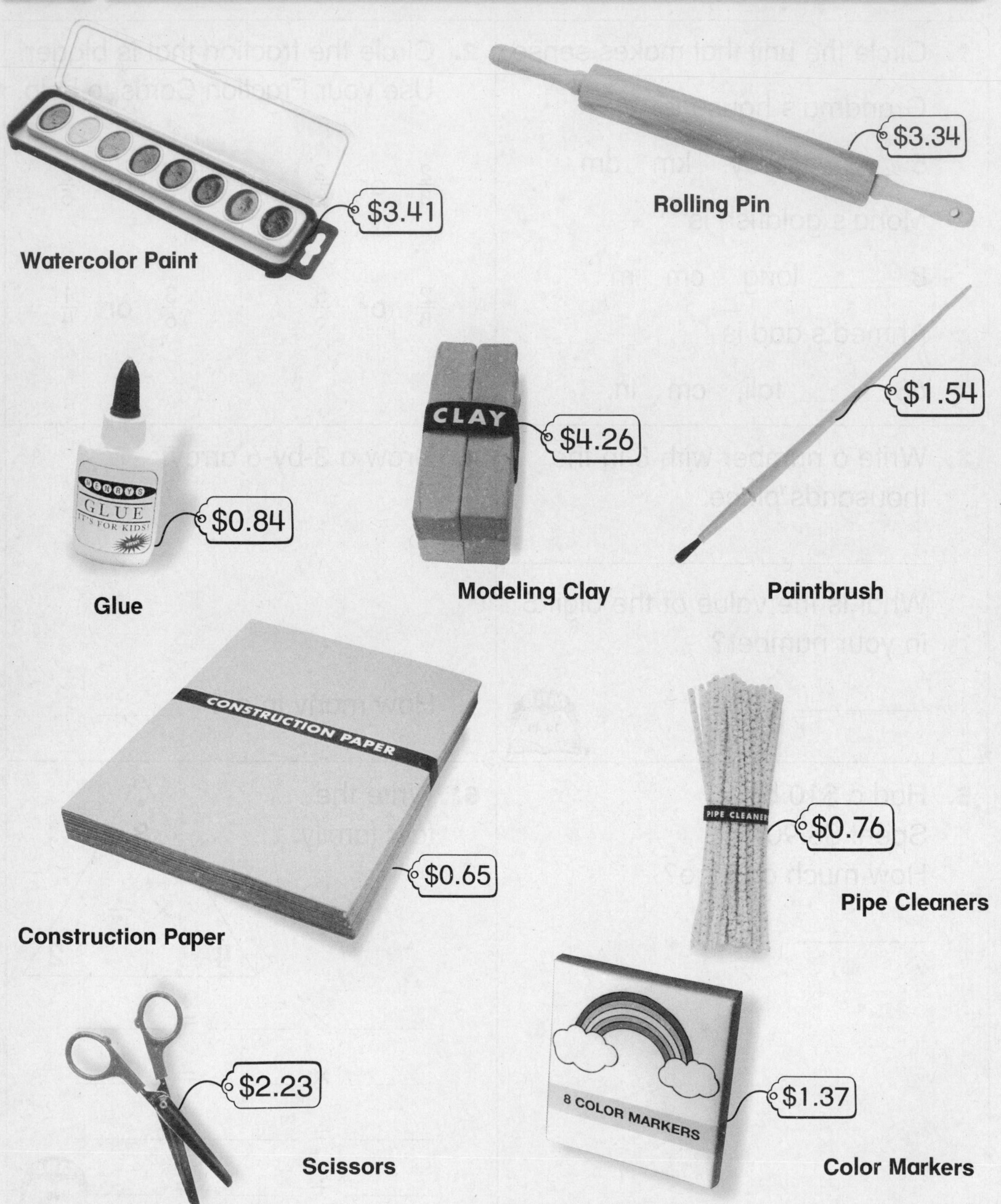

Date Time

Buying Art Supplies

Estimate the total cost for each pair of items.

Write your estimate in the answer space.

Add to find the total cost.

Check your estimate with your total cost.

1. pipe cleaners and watercolors Estimated Cost / Total Cost	**2.** clay and construction paper Estimated Cost / Total Cost	**3.** paintbrush and scissors Estimated Cost / Total Cost
4. glue and construction paper Estimated Cost / Total Cost	**5.** markers and glue Estimated Cost / Total Cost	**6.** clay and rolling pin Estimated Cost / Total Cost

Date Time

LESSON 11·2

Comparing Costs

Use the Art Supply Poster on journal page 264.
In Problems 1–6, circle the item that costs more.
Then find how much more.

1. glue or markers How much more? ______	**2.** construction paper or paintbrush How much more? ______
3. pipe cleaners or paintbrush How much more? ______	**4.** rolling pin or scissors How much more? ______
5. watercolors or markers How much more? ______	**6.** paintbrush or watercolors How much more? ______

7. You buy a pack of construction paper. You pay with a $1 bill.

Should you get more or less than 2 quarters in change? ______

8. You buy pipe cleaners. You pay with a $1 bill.

How much change should you get? ______

9. You buy a rolling pin. You pay with a $5 bill.

How much change should you get? ______

Date Time

LESSON 11·2

Data Analysis

Lily collected information about the ages of people in her family.

Complete.

Name	Age
Dave	40
Dedra	36
Jamal	12
Tyler	10
Lily	8

1. The oldest person is ______________.

Age: ________ years

2. The youngest person is ______________.

Age: ________ years

3. Range of the ages (oldest minus the youngest):

________ years

4. Middle value of the ages: ________ years

Fill in the blanks with the name of the correct person.

5. Jamal is about 4 years older than ______________.

6. Tyler is about 26 years younger than ______________.

7. Dedra is about 3 times as old as ______________.

8. Dave is about 4 times as old as ______________.

Try This

9. Tyler is about $\frac{1}{4}$ as old as ______________.

LESSON 11·2 Math Boxes

1. Solve and show your work.

$$\begin{array}{r} 47 \\ -\ 39 \\ \hline \end{array}$$

Unit
students

2. What is the minimum number (the smallest) in the list?

______

2,371; 429; 578; 1,261

3. What shape is a globe? Circle the best answer.

A. sphere

B. rectangular prism

C. cone

D. cylinder

4. 20 campers divided equally among 5 tents. How many campers in each tent?

______ campers

5. Fill in the missing numbers.

		1,065	
	1,074		

6. What is the range of this set of numbers (the largest minus the smallest)?

______

75, 93, 108, 52

Date Time

LESSON 11·3

Trade-First Subtraction

- Make a ballpark estimate for each problem and write a number model for your ballpark estimate.
- Use the trade-first method of subtraction to solve each problem.

Example:

Ballpark estimate:

40 − 20 = 20

longs 10s	cubes 1s
2	17
~~3~~	~~7~~
− 1	9
1	8

Answer: 18

1. Ballpark estimate:

______

longs 10s	cubes 1s
2	8
− 1	9

Answer

2. Ballpark estimate:

______

longs 10s	cubes 1s
3	1
− 1	7

Answer

3. Ballpark estimate:

______

flats 100s	longs 10s	cubes 1s
1	3	5
−	2	6

Answer

4. Ballpark estimate:

______

flats 100s	longs 10s	cubes 1s
1	4	4
− 1	2	7

Answer

5. Ballpark estimate:

______

flats 100s	longs 10s	cubes 1s
1	2	6
−	4	7

Answer

Date Time

LESSON 11·3

Math Boxes

1. The perimeter is about

_____ cm.

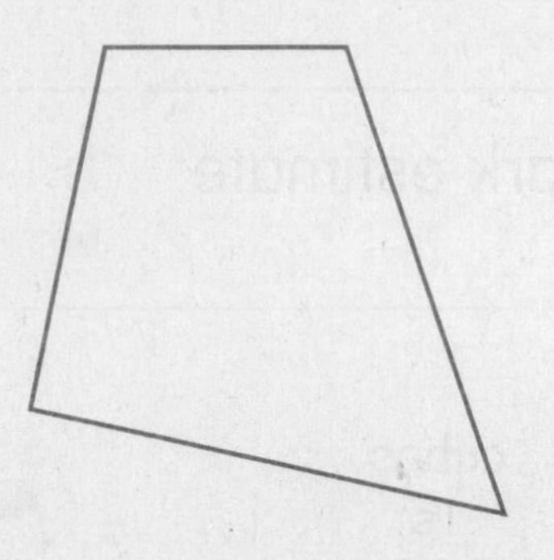

2. Divide into:

halves fourths

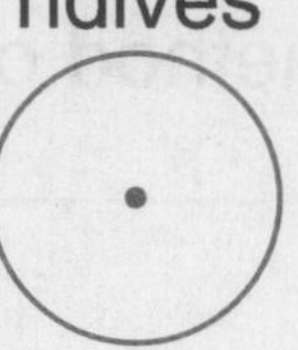

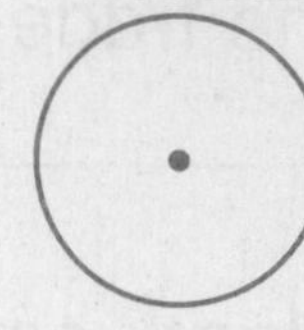

Write <, >, or =.

$\frac{1}{2}$ _____ $\frac{1}{4}$ $\frac{2}{4}$ _____ $\frac{1}{2}$

$\frac{1}{2}$ _____ $\frac{3}{4}$

3. What is the value of the digit 4 in each number?

14 _____

142 _____

436 _____

4,678 _____

4.

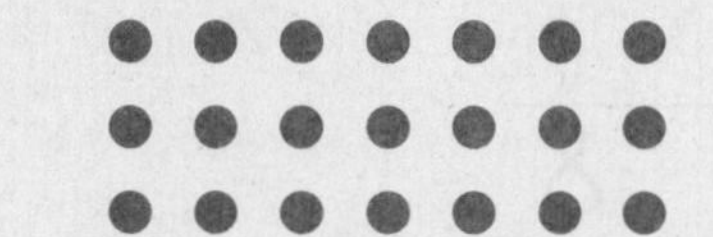

_____-by-_____ array

How many in all? _____

5. I had a 10-dollar bill. I spent \$5.23. How much change did I receive? Fill in the circle next to the best answer.

Ⓐ \$3.80 Ⓑ \$4.77

Ⓒ \$5.00 Ⓓ \$15.23

6. Complete the Fact Triangle. Write the fact family.

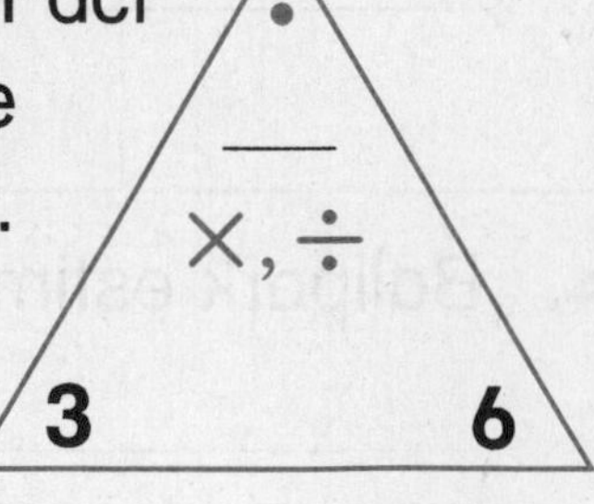

_____ × _____ = _____

_____ × _____ = _____

_____ ÷ _____ = _____

_____ ÷ _____ = _____

LESSON 11·4

Math Boxes

1. Solve and show your work.

$$\begin{array}{r} 71 \\ -\ 23 \\ \hline \end{array}$$

Unit

2. What is the maximum number (the largest) in the list?

____

7,946; 2,599; 17,949; 8,112

3. What shape is a can of soup?

____

4. 15 baseball cards are shared equally among 4 children. How many cards does each child get?

_____ cards

How many left over? _________

5. Fill in the missing numbers.

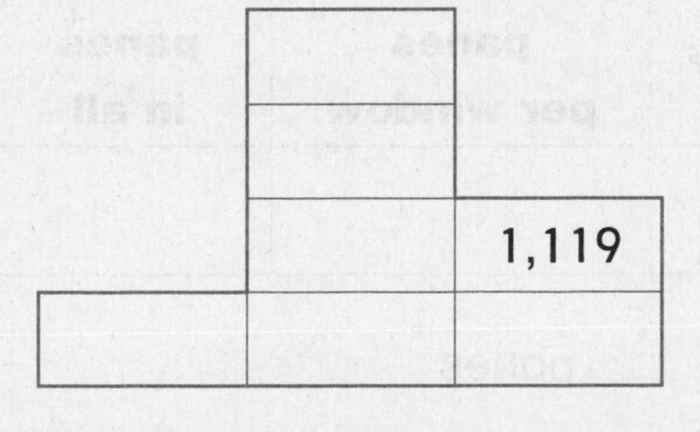

6. What is the range of this list of numbers (the largest minus the smallest)?

____

29, 132, 56, 30

Multiplication Number Stories

1. 4 insects on the flower. How many legs in all?

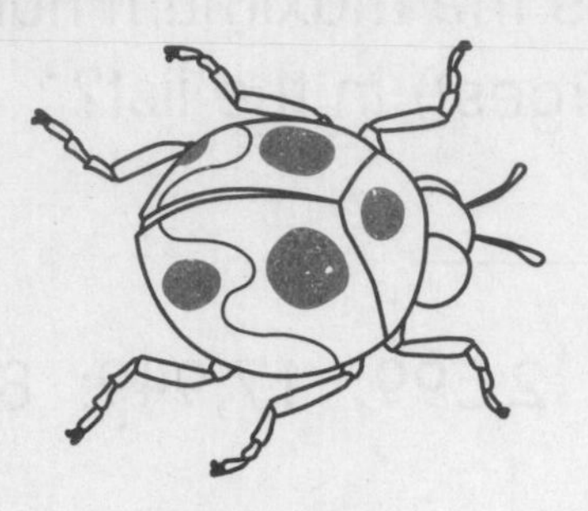

Has 6 legs

insects	legs per insect	legs in all

Answer: _____ legs

Number model: ____________________

2. 3 vans full of people. How many people in all?

Holds 10 people

vans	people per van	people in all

Answer: _____ people

Number model: ____________________

3. 9 windows. How many panes in all?

Has 4 panes

windows	panes per window	panes in all

Answer: _____ panes

Number model: ____________________

Date _______________ Time _______________

Multiplication Number Stories *continued*

Try This

Use the pictures to make up two multiplication number stories.

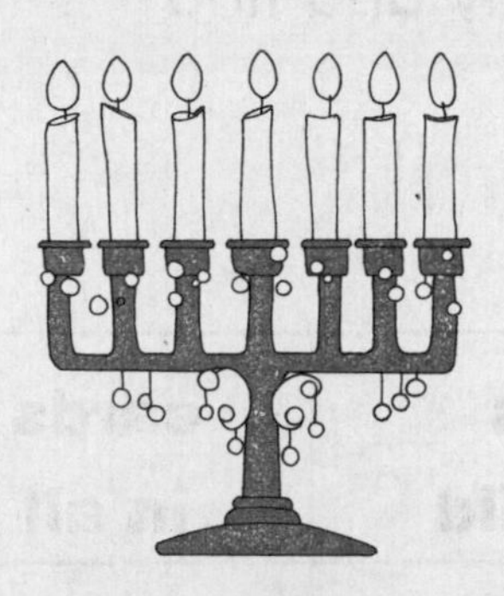

Has 7 candles

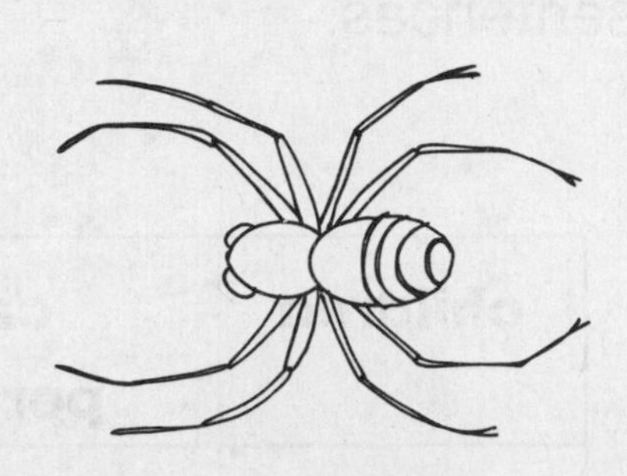

Has 8 legs

Has 5 players

For each story:

- Fill in the multiplication diagram.
- Draw a picture or array and find the answer.
- Fill in the number model.

4. _______________________________

____

____

______	______ per ______	______ in all

Answer: ______

Number model: ______________________

5. _______________________________

____

____

______	______ per ______	______ in all

Answer: ______

Number model: ______________________

LESSON 11·5

Division Number Stories

For each number story:

- Fill in the diagram.
- On a separate sheet of paper, draw a picture or array and find the answer. Complete the sentences.
- Fill in the number model.

1. Five children are playing a game with a deck of 30 cards. How many cards can the dealer give each player?

children	cards per child	cards in all

_____ cards to each player. _____ cards are left over.

Number model: _____ ÷ _____ → _____ R_____

2. The pet shop has 12 puppies in pens. There are 4 puppies in each pen. How many pens have puppies in them?

pens	puppies per pen	puppies in all

_____ pens have puppies in them. _____ puppies are left over.

Number model: _____ ÷ _____ → _____ R_____

3. Tennis balls are sold 3 to a can. Luis buys 15 balls. How many cans is that?

cans	balls per can	balls in all

Luis buys _____ cans.

Number model: _____ ÷ _____ → _____ R_____

Date ______________________ Time ______________________

LESSON 11•5 Division Number Stories *continued*

4. Eight children share 18 toys equally. How many toys does each child get?

______	______ per ______	______ in all

Each child gets ______ toys.

______ toys are left over.

Number model: ______ ÷ ______ → ______ R______

---

5. Seven friends share 24 marbles equally. How many marbles does each friend get?

______	______ per ______	______ in all

Each friend gets ______ marbles. ______ marbles are left over.

Number model: ______ ÷ ______ → ______ R______

Try This

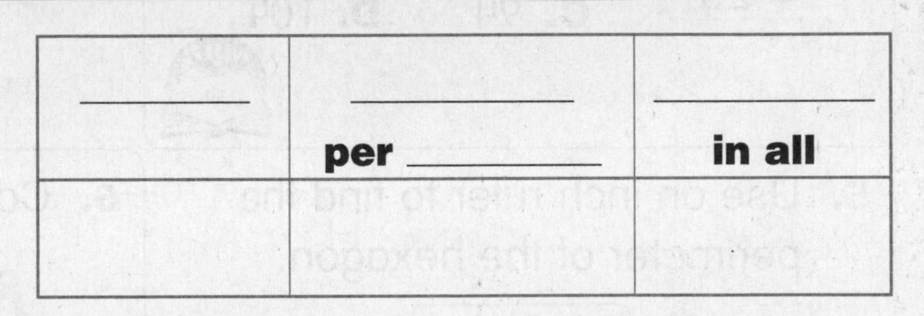

______	______ per ______	______ in all

6. Tina is storing 20 packages of seeds in boxes. Each box holds 6 packages. How many boxes does Tina need to store all the packages? (Be careful. Think!)

Tina needs ______ boxes.

Number model: ______ ÷ ______ → ______ R______

Date Time

LESSON 11•5

Math Boxes

1. 8 flower boxes. 4 plants in each box. How many plants?

_____ plants

Fill in the diagram and write a number model.

boxes	plants per box	plants in all

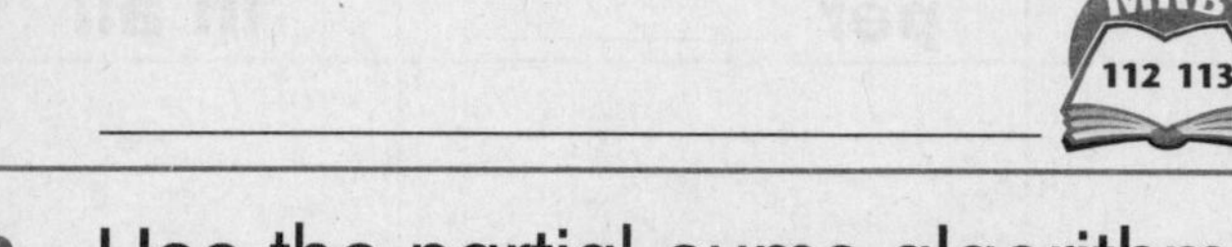

____

MRB 112 113

2. ●●●●
●●●●

_____-by-_____ array

How many dots in all? _____

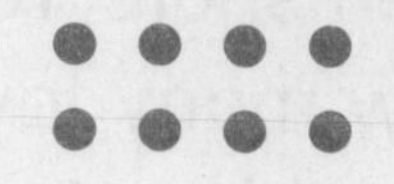

3. Use the partial-sums algorithm to solve. Show your work. Circle the best answer.

$$\begin{array}{r} 78 \\ +\ 26 \\ \hline \end{array}$$

A. 92 **B.** 52

C. 94 **D.** 104

MRB 30

4. Count by ones.

5,099; _____; _____;

5,102; _____; _____;

5,105; _____; _____

5. Use an inch ruler to find the perimeter of the hexagon.

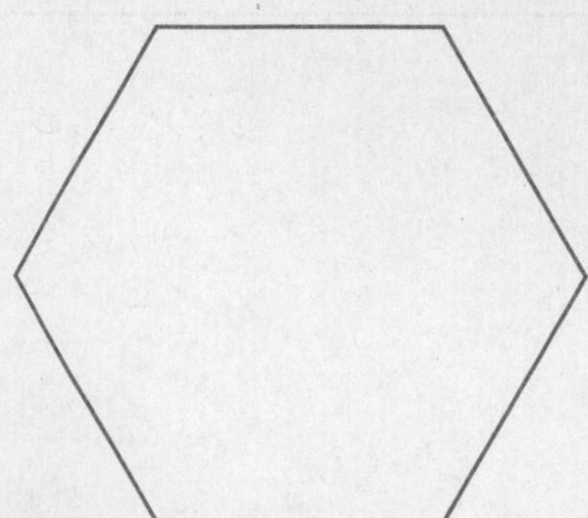

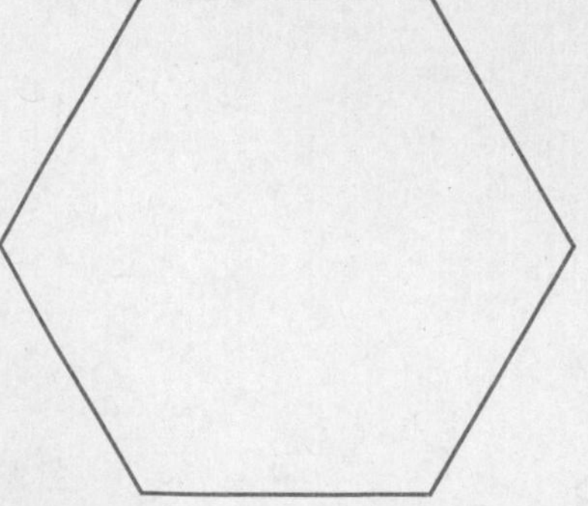

The perimeter is _____ inches.

MRB 68

6. Complete the table.

Rule

×2

in	out
0	
1	
2	
	6
4	
	10

MRB 100–102

Date Time

LESSON 11•6

Multiplication Facts List

I am listing the times _____ facts.
If you are not sure of a fact, draw an array with Os or Xs.

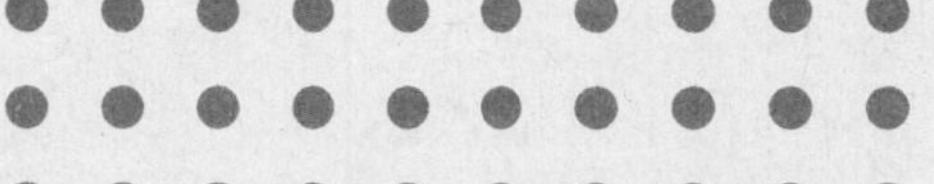

2 × _____ = _____

3 × _____ = _____

4 × _____ = _____

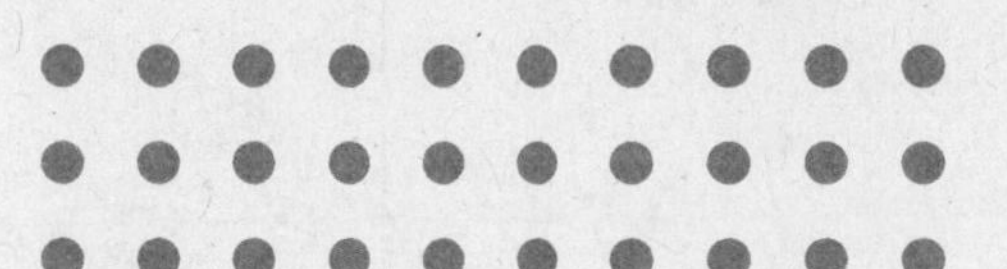

5 × _____ = _____

6 × _____ = _____

7 × _____ = _____

8 × _____ = _____

9 × _____ = _____

10 × _____ = _____

Date Time

LESSON 11·6

Using Arrays to Find Products

Draw an array to help you find each product.
Use Xs to draw your arrays.

1. $2 \times 4 =$ ______ X X X X X X X X	**2.** $4 \times 2 =$ ______
3. $6 \times 5 =$ ______	**4.** $5 \times 6 =$ ______
5. $5 \times 5 =$ ______	**6.** $2 \times 10 =$ ______

Try This

7. $4 \times 15 =$ ______

Date Time

LESSON 11·6

Math Boxes

1. Subtract. Show your work.

$$\begin{array}{r} 72 \\ -\ 35 \\ \hline \end{array}$$

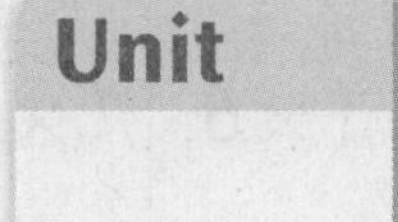

2. What is the median (the middle value) for this list of numbers?

_____

51, 82, 51, 23, 23, 67

3. This line segment is

_____ cm long.

____

Draw a line segment 4 cm longer.

4. I have a pile of 16 counters.

$\frac{1}{2}$ = _____ counters

$\frac{8}{16}$ = _____ counters

5. Find the rules.

6. Write the mode for this set of numbers (the number that occurs the most often).

_____

29, 17, 39, 12, 17

MRB 45

LESSON 11·7

Products Table

0×0 = **0**	0×1 =	0×2 =	0×3 =	0×4 =	0×5 =	0×6 =	0×7 =	0×8 =	0×9 =	0×10 =
1×0 =	1×1 = **1**	1×2 =	1×3 =	1×4 =	1×5 =	1×6 =	1×7 =	1×8 =	1×9 =	1×10 =
2×0 =	2×1 =	2×2 = **4**	2×3 =	2×4 =	2×5 =	2×6 =	2×7 =	2×8 =	2×9 =	2×10 =
3×0 =	3×1 =	3×2 =	3×3 = **9**	3×4 =	3×5 =	3×6 =	3×7 =	3×8 =	3×9 =	3×10 =
4×0 =	4×1 =	4×2 =	4×3 =	4×4 = **16**	4×5 =	4×6 =	4×7 =	4×8 =	4×9 =	4×10 =
5×0 =	5×1 =	5×2 =	5×3 =	5×4 =	5×5 = **25**	5×6 =	5×7 =	5×8 =	5×9 =	5×10 =
6×0 =	6×1 =	6×2 =	6×3 =	6×4 =	6×5 =	6×6 = **36**	6×7 =	6×8 =	6×9 =	6×10 =
7×0 =	7×1 =	7×2 =	7×3 =	7×4 =	7×5 =	7×6 =	7×7 = **49**	7×8 =	7×9 =	7×10 =
8×0 =	8×1 =	8×2 =	8×3 =	8×4 =	8×5 =	8×6 =	8×7 =	8×8 = **64**	8×9 =	8×10 =
9×0 =	9×1 =	9×2 =	9×3 =	9×4 =	9×5 =	9×6 =	9×7 =	9×8 =	9×9 = **81**	9×10 =
10×0 =	10×1 =	10×2 =	10×3 =	10×4 =	10×5 =	10×6 =	10×7 =	10×8 =	10×9 =	10×10 = **100**

LESSON 11·7

Math Boxes

1. 5 nests with 3 eggs in each. How many eggs in all?

_____ eggs

nests	eggs per nest	eggs in all

____

MRB 112 113

2. Maria has 9 pairs of shoes in her closet. How many shoes does she have in all?

Draw an array.

____

_____ shoes

3. Solve.

Unit: baby alligators

_____ = 24 + 41

33 + 12 = _____

_____ = 52 + 15

16 + 51 = _____

4. Count by thousands. Fill in the circle next to the best answer.

2,324; _______; 4,324

Ⓐ 3,224 Ⓑ 2,424

Ⓒ 3,324 Ⓓ 3,434

5. Draw a rectangle.
Make two sides 3 inches long.
Make the other two sides 2 inches long.

6. Complete the table.

Rule: ×2

in	out
20	
	60
40	

MRB 100–102

LESSON 11·8

Multiplication/Division Fact Families

Write the fact family for each Fact Triangle.

1.

10
×, ÷
2 5

5 × 2 = 10

___ × ___ = ___

10 ÷ 2 = 5

___ ÷ ___ = ___

2.

12
×, ÷
3 4

___ × ___ = ___

___ × ___ = ___

___ ÷ ___ = ___

___ ÷ ___ = ___

3.

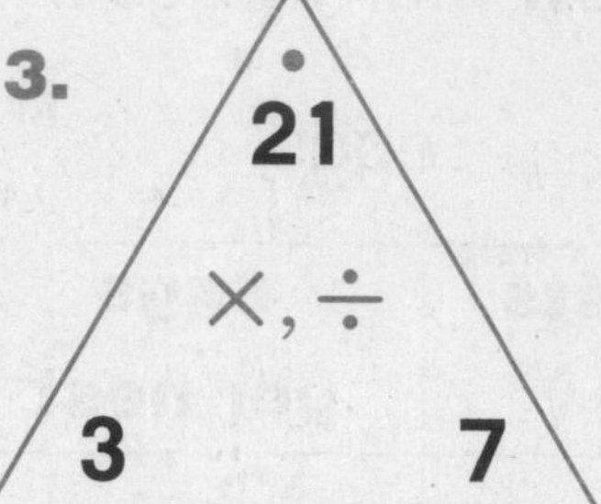

___ × ___ = ___

___ × ___ = ___

___ ÷ ___ = ___

___ ÷ ___ = ___

4.

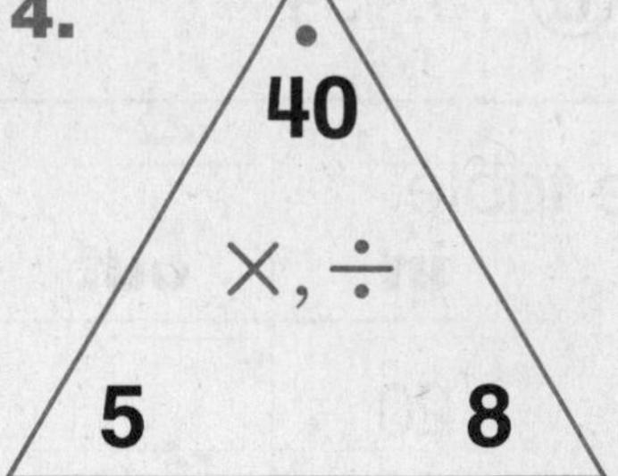

___ × ___ = ___

___ × ___ = ___

___ ÷ ___ = ___

___ ÷ ___ = ___

5.

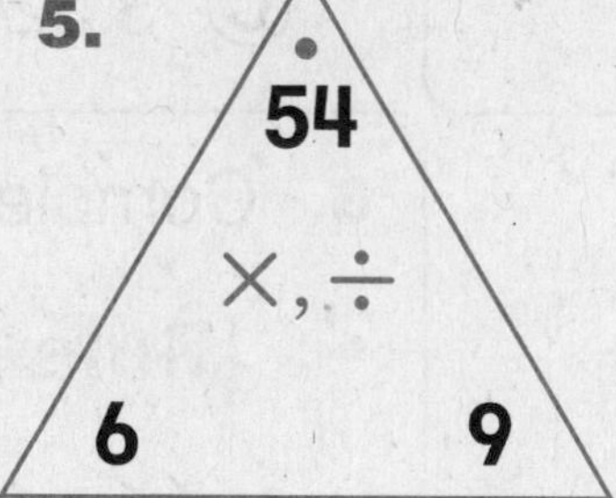

___ × ___ = ___

___ × ___ = ___

___ ÷ ___ = ___

___ ÷ ___ = ___

6.

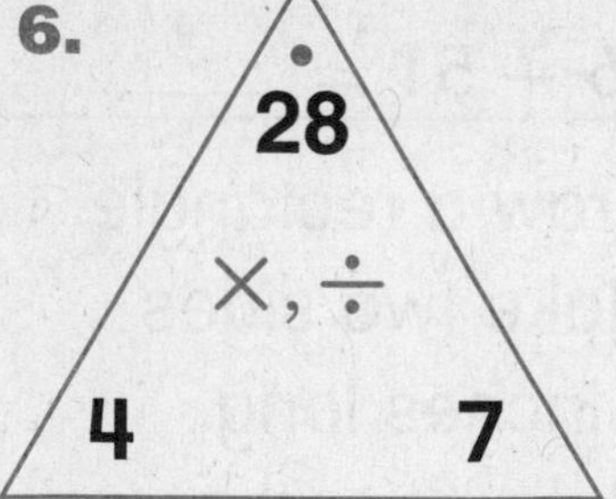

___ × ___ = ___

___ × ___ = ___

___ ÷ ___ = ___

___ ÷ ___ = ___

LESSON 11·8

Multiplication and Division with 2, 5, and 10

1. double 8 = _____ _____ = double 9

$2 \times 4 =$ _____ _____ $= 2 \times 1$

$7 \times 2 =$ _____ _____ $= 0 \times 2$

2. 40 cents = _____ nickels $5 \div 1 =$ _____

$15 \div 5 =$ _____ _____ nickels = 25 cents

3. 40 cents = _____ dimes _____ $\div 10 = 6$

$20 \div 10 =$ _____ _____ dimes = 90 cents

4. For each multiplication fact, give two division facts in the same fact family.

$2 \times 6 = 12$ 12 ÷ 2 = 6 12 ÷ 6 = 2

$5 \times 9 = 45$ _____ ÷ _____ = _____ _____ ÷ _____ = _____

$10 \times 4 = 40$ _____ ÷ _____ = _____ _____ ÷ _____ = _____

$3 \times 2 = 6$ _____ ÷ _____ = _____ _____ ÷ _____ = _____

$8 \times 5 = 40$ _____ ÷ _____ = _____ _____ ÷ _____ = _____

$5 \times 10 = 50$ _____ ÷ _____ = _____ _____ ÷ _____ = _____

Date Time

LESSON 11·8

Math Boxes

1. Subtract. Show your work.

90	37
– 64	– 18

Unit

2. What is the median (the middle value) for this list of numbers?

______

50, 31, 41, 42, 41

3. Draw a line segment 3 cm long.

Draw a second line segment 4 cm longer than the first.

Draw a third line segment twice as long as the first.

4. Get 21 counters.

$\frac{1}{3}$ = ______ counters

$\frac{2}{7}$ = ______ counters

$\frac{3}{3}$ = ______ counters

5. Complete the frames.

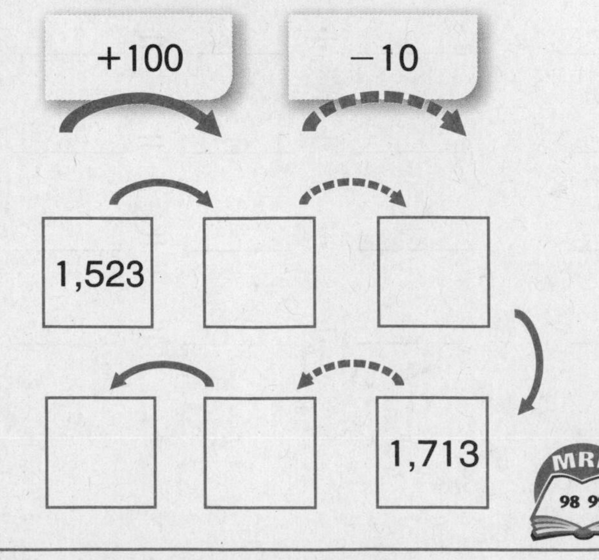

6. Find the mode (the number that occurs most often). Choose the best answer.

496, 738, 713, 100, 713

⬭ 713 ⬭ 496

⬭ 100 ⬭ 738

MRB 45

Date Time

LESSON 11·9

Math Boxes

1. Each ladybug has 5 spots. 20 spots in all.

How many ladybugs? ______

Fill in the diagram and write a number model.

ladybugs	spots per ladybug	spots in all

MRB 112 113

______

2. Multiply. If you need help, make arrays.

Unit

$4 \times 5 =$ ______

$6 \times 2 =$ ______

$1 \times 10 =$ ______

3. Complete.

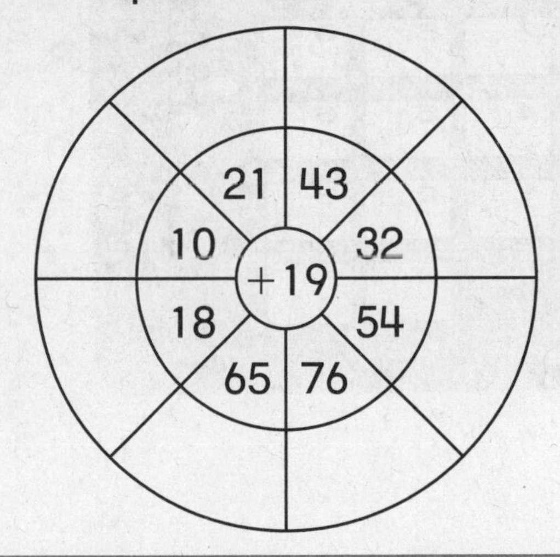

4. Write the number that is 1,000 more than:

7,542 ______

1,837 ______

5,641 ______

8,863 ______

5. Draw a line 1 inch long.

Draw another line twice as long.

6. Find the rule. Complete the table.

in	out
10¢	20¢
15¢	30¢
	50¢
60¢	

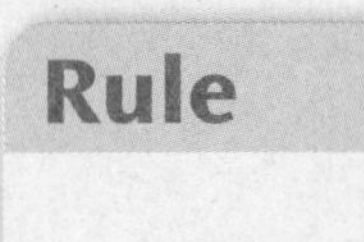

MRB 100–102

Date Time

LESSON 11·9

Beat the Calculator

Materials ☐ calculator

Players 3 (Caller, Brain, and Calculator)

Directions

1. The Caller reads fact problems from the Brain's journal—in the order listed on the next page.
2. The Brain solves each problem and says the answer.
3. While the Brain is working on the answer, the Calculator solves each problem using a calculator and says the answer.
4. If the Brain beats the Calculator, the Caller makes a check mark next to the fact in the Brain's journal.

Date Time

Beat the Calculator *continued*

✓	✓	✓	Fact Problem
			2 × 4 = ____
			3 × 5 = ____
			2 × 2 = ____
			4 × 3 = ____
			5 × 5 = ____
			6 × 2 = ____
			6 × 5 = ____
			3 × 3 = ____
			4 × 5 = ____
			3 × 6 = ____

✓	✓	✓	Fact Problem
			7 × 3 = ____
			5 × 2 = ____
			6 × 4 = ____
			2 × 7 = ____
			3 × 2 = ____
			4 × 4 = ____
			4 × 1 = ____
			4 × 7 = ____
			7 × 5 = ____
			0 × 2 = ____

LESSON 11·10

Math Boxes

1. What is the range of this set of numbers (the largest number minus the smallest number)?

____

81, 910, 109, 175

2. Find the mode (the number that occurs most often).

____

183, 56, 618, 56,
215, 56, 183, 56

3. Complete the Fact Triangle. Write the fact family.

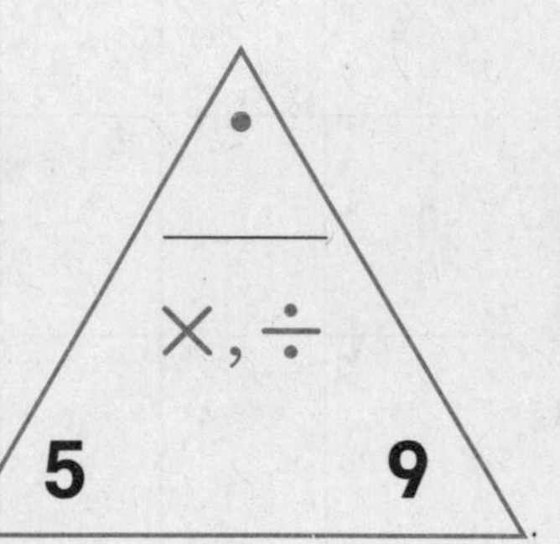

_______ × _______ = _______

_______ × _______ = _______

_______ ÷ _______ = _______

_______ ÷ _______ = _______

4. Find the median (the middle number).

____

640, 710, 615, 915, 320

5. Complete the table.

Rule
×2

in	out
100	
	400
1,000	
	4,000

6. Complete the Fact Triangle. Write the fact family.

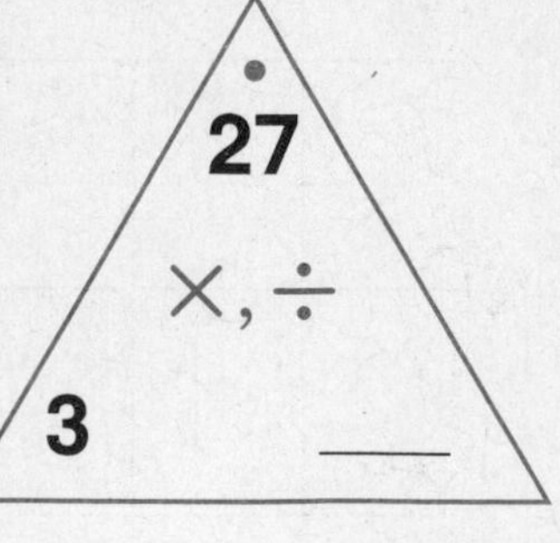

_______ × _______ = _______

_______ × _______ = _______

_______ ÷ _______ = _______

_______ ÷ _______ = _______

MRB 38

LESSON 12•1 Review: Telling Time

1. How many hours do clock faces show? _____ hours
2. How long does it take the hour hand to move from one number to the next? ______________________________
3. How long does it take the minute hand to move from one number to the next? ______________________________
4. How many times does the hour hand move around the clock face in one day? _____ times
5. How many times does the minute hand move around the clock face in one day? _____ times

Write the time shown by each clock.

6. 

_____ : _____

7. 

_____ : _____

8. 

_____ : _____

Draw the hour and minute hands to match the time.

9. 

8:00

10. 

6:45

11. 

4:10

Date Time

LESSON 12·1

Math Boxes

1. Jordan spent \$6.38 on a book and \$1.23 on a magazine. How much did he spend all together? First estimate the costs and the total.

______ + ______ = ______

Then use partial sums and solve.

2. 1 hour = ____ minutes

$\frac{1}{2}$ hour = ____ minutes

$\frac{1}{4}$ hour = ____ minutes

$\frac{3}{4}$ hour = ____ minutes

$1\frac{1}{2}$ hours = ____ minutes

3. Write a fraction for each shaded part. Put <, >, or = in the box.

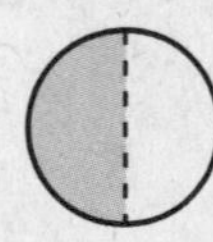

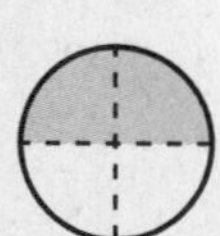

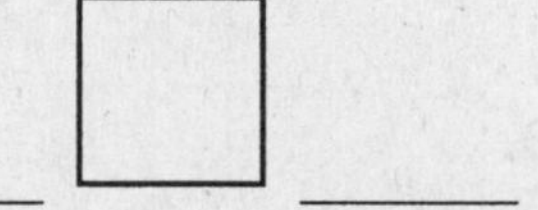

4. I spent \$4.22 at the store and gave the cashier a \$10 bill. How much change should I get?

\$______

5.

5,401	1,290	632
3,679	890	798

The minimum number is ______.

The maximum number is ______.

6. A pentagon has ___ sides.

A hexagon has ___ sides.

An octagon has ___ sides.

Date Time

LESSON 12-2

Time Before and After

1. It is:

Show the time 20 minutes later.

What time is it?

_____ : _____

2. It is:

Show the time 35 minutes later.

What time is it?

_____ : _____

3. It is:

Show the time 15 minutes earlier.

What time is it?

_____ : _____

4. You pick a time. Draw the hands on the clock.

It is:

Show the time 50 minutes later.

What time is it?

_____ : _____

Date Time

LESSON 12·2

Many Names for Times

What time does each clock show? Fill in the ovals next to the correct names.

Example: ⬮ quarter-past 1 ⬮ 15 minutes after 1

O two ten O 5 minutes after 3

⬮ one fifteen

1. O seven fifteen O quarter-to 7
   O quarter-to 8 O quarter-past 8
   O quarter-past 7

2. O half-past 10 O eleven thirty
   O half-past 11 O 30 minutes after 10
   O ten thirty

3. O quarter-past 5 O quarter-to 6
   O quarter-to 5 O six fifteen
   O five forty-five

4. O nine forty O 20 to 8
   O 20 to 9 O eight forty
   O 40 minutes to 9

Date Time

LESSON 12•2

Addition and Subtraction Strategies

Add or subtract. Use your favorite addition or subtraction strategy.

1. 53 + 45 Answer	**2.** 36 + 48 Answer	**3.** 456 + 17 Answer
4. 68 − 24 Answer	**5.** 65 − 27 Answer	**6.** 516 − 38 Answer

LESSON 12•2

Math Boxes

1. **Favorite Animals of Ms. Lee's Classroom**

Whales	XXX
Manatees	XXXXXX
Walruses	XXXXX
Cubs	XXXXXXXXXX

Which animal was the most popular? ______

2. Solve.

Unit

$(14 - 7) + 4 =$ ____

$14 - (7 + 4) =$ ____

$(12 - 7) + 5 =$ ____

$12 - (7 + 5) =$ ____

3. Write < or >.

2,469 _____ 12,469

60,278 _____ 50,278

25,100 _____ 25,110

MRB 9

4. What item is the shape of a rectangular prism? Circle the best answer.

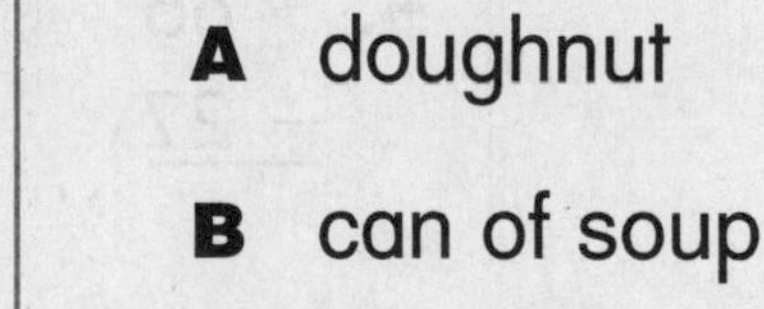

A doughnut

B can of soup

C book

D ice-cream cone

5.

The area is _____ sq cm.

The perimeter is _____ cm.

6. The pet store sold 12 fish. $\frac{1}{2}$ were guppies and $\frac{1}{4}$ were neons. The rest were angelfish. How many of each?

There were _____ guppies.

There were _____ neons.

There were _____ angelfish.

Date Time

Important Events in Communication

For each event below, make a dot on the timeline and write the letter for the event above the dot.

A telephone (1876)

B radio (1906)

C television (1926)

D telegraph (1837)

E CD player (1982)

F copier (1937)

G audiocassette (1963)

H phonograph (1877)

I personal computer (1974)

J movie machine (1894)

K 3-D movies (1922)

L videocassette (1969)

M typewriter (1867)

N FM radio (1933)

A

1830 1840 1850 1860 1870 1880 1890 1900 1910 1920 1930 1940 1950 1960 1970 1980 1990 2000 2010

Date Time

LESSON 12•3

Math Boxes

1. A baseball costs \$3.69. A yo-yo costs \$1.49. You buy both.

   Estimate the cost:

   ________ + ________ = ________

   Actual cost:

   \$________

2. Naquon grew 2 inches in ______. Circle the best answer.

   A 1 day

   B 5 minutes

   C 1 year

   D 24 hours

3. Write four names for $\frac{1}{2}$. Use your Fraction Cards to help.

   ______, ______, ______, ______

   Write 2 fractions that are:

   greater than $\frac{1}{2}$. ______, ______

   less than $\frac{1}{2}$. ______, ______

4. I bought a beach ball for \$1.49 and a sand toy for \$3.96. How much change will I get from a \$10 bill?

   \$________

5. \$24.06 \$9.99 \$14.98 \$19.99 \$29.83

   The maximum is

   \$________.

   The minimum is

   \$________.

6. Draw two polygons with 4 sides.

Date Time

LESSON 12•4

Subtraction Practice

Fill in the unit box. Then, for each problem:

Unit

- Make a ballpark estimate before you subtract.
- Write a number model for your estimate. Then solve the problem. For Problems 1 and 2, use the trade-first algorithm. For Problems 3–6, use any strategy you choose.
- Compare your estimate to your answer.

1. Ballpark estimate: ______ 25 – 18 = ______	**2.** Ballpark estimate: ______ 31 – 22 = ______	**3.** Ballpark estimate: ______ 53 – 29 = ______
4. Ballpark estimate: ______ 87 – 39 = ______	**5.** Ballpark estimate: ______ 148 – 29 = ______	**6.** Ballpark estimate: ______ 177 – 48 = ______

Date Time

LESSON 12·4

Math Boxes

1\.

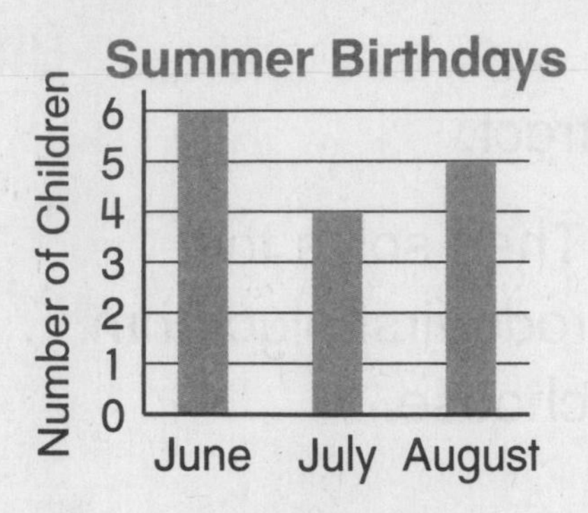

How many birthdays are in June and July? ______

2\. Add parentheses to make the number models true.

Unit
children

18 – 13 – 4 = 9

18 – 13 – 4 = 1

27 – 6 + 10 = 31

4 × 2 + 3 = 20

3\. Write <, >, or =.

10,000 ______ 9,999

33,231 ______ 30,231

75,679 ______ 75,855

4\. Name 3 objects that are shaped like a cone.

____

____

____

5\. Solve.

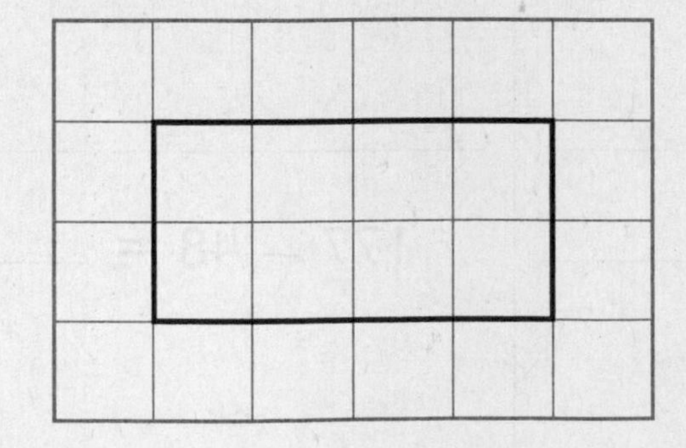

Area: ______ sq cm

Perimeter: ______ cm

6\. Mrs. Bell had 30 pennies. She gave $\frac{1}{3}$ of the pennies to Max and $\frac{1}{2}$ of the pennies to Julie.

Max received ______ pennies.

Julie received ______ pennies.

How many pennies did Mrs. Bell have left?

______ pennies

Date Time

LESSON 12·5

Related Multiplication and Division Facts

Solve each multiplication fact. Use the fact triangles to help you.

Then use the three numbers to write two division facts.

1. $3 \times 7 =$ 21

21 ÷ 7 = 3

21 ÷ 3 = 7

2. $3 \times 8 =$ ____

____ ÷ ____ = ____

____ ÷ ____ = ____

3. $3 \times 9 =$ ____

____ ÷ ____ = ____

____ ÷ ____ = ____

4. $4 \times 7 =$ ____

____ ÷ ____ = ____

____ ÷ ____ = ____

5. $4 \times 8 =$ ____

____ ÷ ____ = ____

____ ÷ ____ = ____

6. $4 \times 9 =$ ____

____ ÷ ____ = ____

____ ÷ ____ = ____

7. $5 \times 7 =$ ____

____ ÷ ____ = ____

____ ÷ ____ = ____

8. $5 \times 8 =$ ____

____ ÷ ____ = ____

____ ÷ ____ = ____

LESSON 12·5

Addition Card Draw Directions

Materials
- ☐ score sheet from *Math Masters,* p. 446
- ☐ 4 each of number cards 1–10
- ☐ 1 each of the number cards 11–20
- ☐ slate or scratch paper

Players 2

Skill Add 3 numbers

Object of the Game To get the higher total

Directions

Shuffle the cards and place the deck with the numbers facing down. Take turns.

1. Draw the top 3 cards from the deck.
2. Record the numbers on the score sheet. Put the 3 cards in a separate pile.
3. Find the sum. Use your slate or paper to do the computation.

After 3 turns:

4. Check your partner's work. Use a calculator.
5. Find the total of the 3 answers. Write the total on the score sheet. The player with the higher total wins.

Date ____________ Time ____________

LESSON 12·5

Math Boxes

1. Write the number that is

	10 more	100 less
368	______	______
4,789	______	______
40,870	______	______
1,999	______	______

2. How many days per week?

______

How many minutes per hour?

______

How many hours per day?

______

How many weeks per year?

______

3. Shade $\frac{1}{2}$ of the shape. Write the equivalent fraction.

______

4. Fill in the table.

in	out
1	
6	
10	
	80
	250

Rule
×10

5. The time is

____ : ____.

20 minutes later will be

____ : ____.

15 minutes earlier was

____ : ____.

6. Solve.

Unit

____ − 23 = 17

60 − ____ = 28

49 = ____ − 21

54 = 80 − ____

Animal Bar Graph

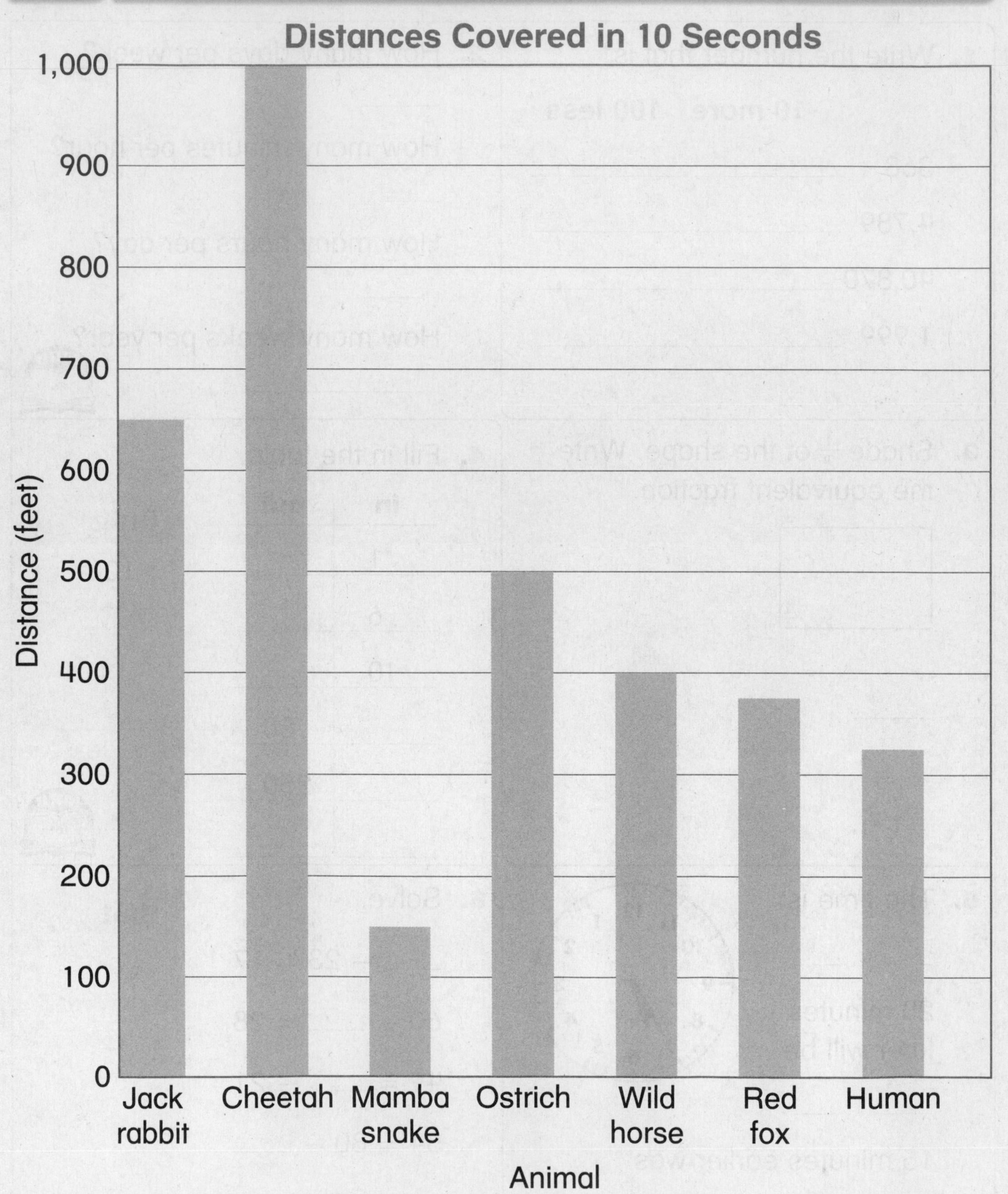

Date Time

LESSON 12•6

Interpreting an Animal Bar Graph

1. In the table, list the animals in order of distance covered in 10 seconds. List the animals from the greatest distance to the least distance.

Distances Covered in 10 Seconds

Animal	Distance
greatest: ______	______ ft
______	______ ft
______	______ ft
______	______ ft
______	______ ft
______	______ ft
least: ______	______ ft

2. Find the middle value of the distances. The middle value is also called the **median.**

   The median is ______ feet.

3. The longest distance is

   ______ feet.

   The shortest distance is

   ______ feet.

4. Fill in the comparison diagram with the longest distance and the shortest distance.

Quantity

Quantity	
	______ Difference

5. Find the difference between the longest and shortest distances. The difference between the largest and smallest numbers in a data set is called the **range.**

   The range is ______ feet.

Date Time

LESSON 12•6

Math Boxes

1. Complete the bar graph.

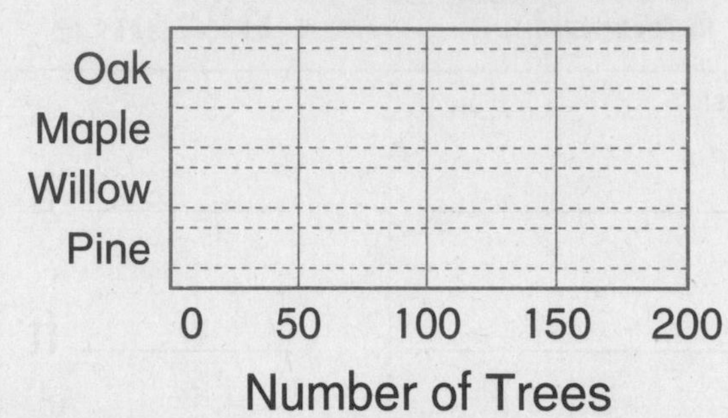

Oak: 100 Maple: 200

Willow: 50 Pine: 150

2. Put parentheses to make each number model true.

$21 = 39 - 10 - 8$

$4 \times 3 + 7 = 40$

$3 \times 5 + 2 = 17$

3. Write <, >, or =.

20,739 ______ 24,596

10,670 ______ 6,670

15,139 ______ 15,264

4. Name the shape.

basketball ____________

shoe box ____________

paper towel roll ____________

5. Use the tick marks to draw lines to show square units. Find the area.

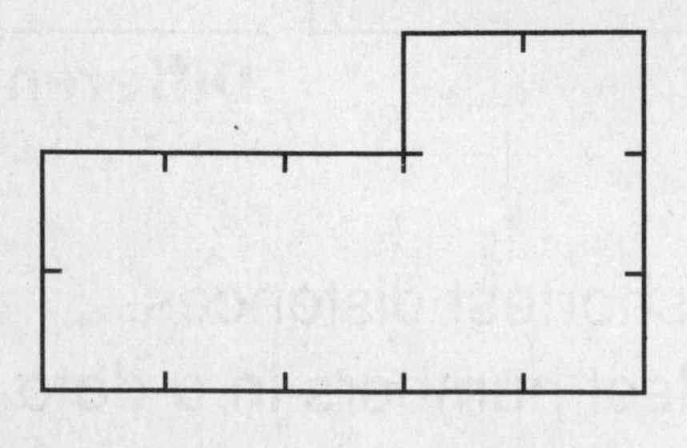

Area ______ sq cm

MRB 69

6. A shark swam 80 miles. A seal swam $\frac{1}{2}$ as far as the shark.

How far did the seal swim?

______ miles

A dolphin swam twice as far as the shark. How far did the dolphin swim?

______ miles

Date Time

LESSON 12•7

Height Changes

The data in the table show the height of 30 children at ages 7 and 8. Your teacher will show you how to make a line plot for the data.

Student	Height	
	7 Years	8 Years
#1	120 cm	123 cm
#2	132 cm	141 cm
#3	112 cm	115 cm
#4	122 cm	126 cm
#5	118 cm	122 cm
#6	136 cm	144 cm
#7	123 cm	127 cm
#8	127 cm	133 cm
#9	115 cm	120 cm
#10	119 cm	125 cm
#11	122 cm	126 cm
#12	103 cm	107 cm
#13	129 cm	136 cm
#14	124 cm	129 cm
#15	109 cm	110 cm

Student	Height	
	7 Years	8 Years
#16	118 cm	122 cm
#17	120 cm	126 cm
#18	141 cm	148 cm
#19	122 cm	127 cm
#20	120 cm	126 cm
#21	120 cm	124 cm
#22	136 cm	142 cm
#23	115 cm	118 cm
#24	122 cm	130 cm
#25	124 cm	129 cm
#26	123 cm	127 cm
#27	131 cm	138 cm
#28	126 cm	132 cm
#29	121 cm	123 cm
#30	118 cm	123 cm

LESSON 12·7

Height Changes *continued*

Use the line plot your class made to make a frequency table for the data.

Frequency Table	
Change in Height	**Number of Children**
0 cm	
1 cm	
2 cm	
3 cm	
4 cm	
5 cm	
6 cm	
7 cm	
8 cm	
9 cm	
10 cm	

Date Time

LESSON 12•7

Height Changes *continued*

1. Make a bar graph of the data in the frequency table.

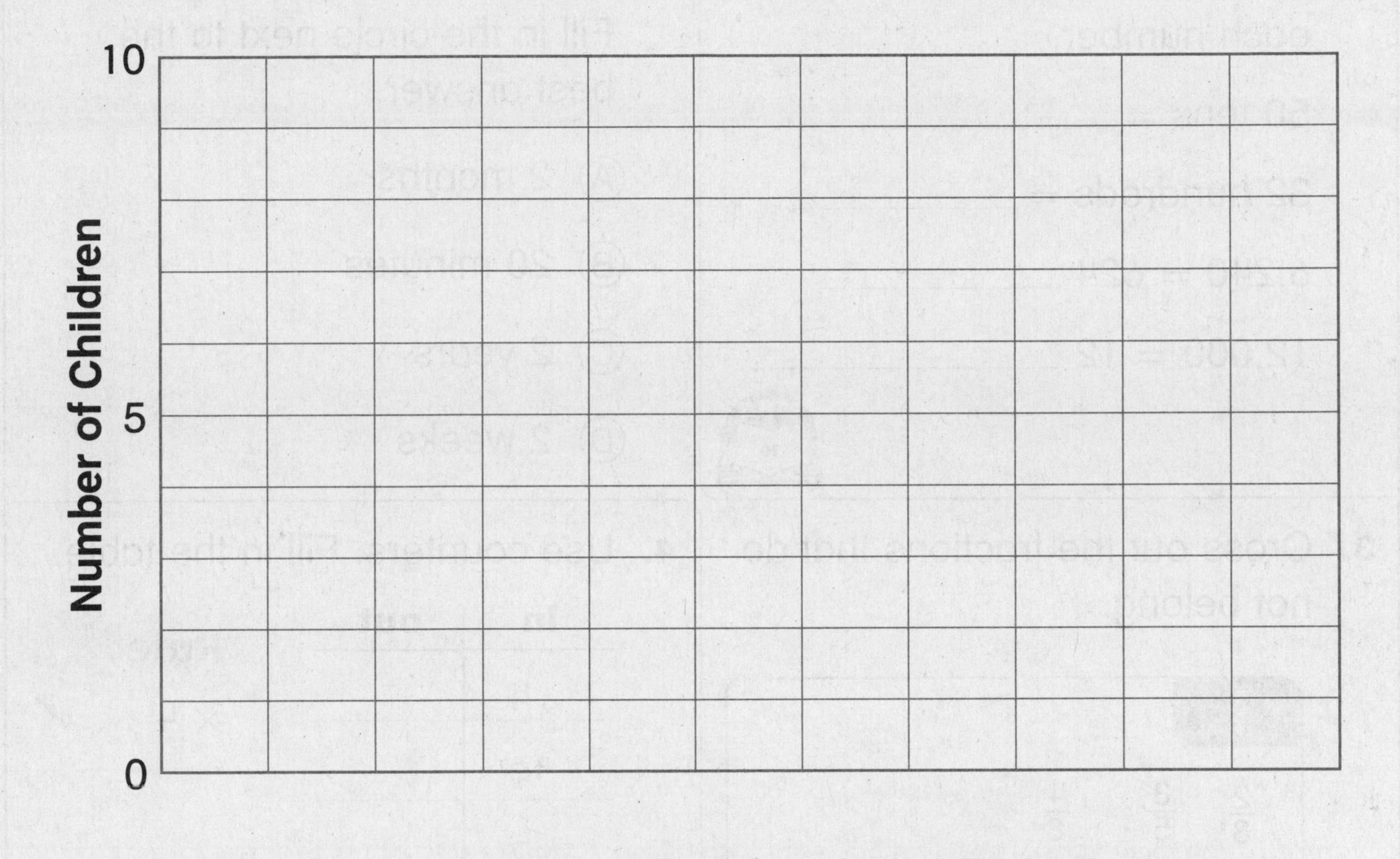

2. The minimum is _____ centimeter(s).

3. The maximum is _____ centimeter(s).

4. The median (the middle value) for the height change data is _____ centimeter(s).

5. The mode (the height change that occurred most often) is _____ centimeter(s).

6. The range is _____ centimeter(s).

Date Time

LESSON 12•7 Math Boxes

1. Write another name for each number.

50 tens = ________

32 hundreds = ________

6,240 = 624 ________

12,000 = 12 ________

2. Jim ate dinner in ________. Fill in the circle next to the best answer.

Ⓐ 2 months

Ⓑ 20 minutes

Ⓒ 2 years

Ⓓ 2 weeks

3. Cross out the fractions that do not belong.

$\frac{1}{2}$

$\frac{2}{3}$, $\frac{3}{5}$, $\frac{4}{8}$,

$\frac{6}{12}$, $\frac{6}{8}$, $\frac{5}{10}$, $\frac{1}{4}$

4. Use counters. Fill in the table.

in	out
4	
9	
7	
	20
	40

Rule

× 4

5. Write the time in hours and minutes.

10 minutes past 12 ____:____

quarter to 11 ____:____

half-past 7 ____:____

25 minutes to 8 ____:____

6. Trade first. Then subtract.

$$\begin{array}{r} \$5.44 \\ -\ \$0.29 \\ \hline \end{array} \qquad \begin{array}{r} \$5.44 \\ -\ \$3.29 \\ \hline \end{array}$$

MRB 34 35

LESSON 12·8

Math Boxes

1. Write the number. Use your Place-Value Book if you need help.

3 tens = ______

33 tens = ______

333 tens = ______

2. Match.

1 day	14 days
3 days	48 hours
2 weeks	24 hours
2 days	72 hours

3. Write the fractions.

______ or ______

4. Fill in the table.

Rule

×3

in	out
0	
1	
2	
3	
	12
	30

5. Cross out names that don't belong.

6:15

six fifteen, quarter to 7,

quarter past 6,

15 minutes before 6,

15 minutes after 6

6. Solve.

Unit

$$\begin{array}{r} 687 \\ -\ 409 \\ \hline \end{array} \qquad \begin{array}{r} 569 \\ -\ 372 \\ \hline \end{array}$$

Date Time

Table of Equivalencies

Weight	
kilogram	1,000 g
pound	16 oz
ton	2,000 lb
1 ounce is about 30 g	

Length	
kilometer	1,000 m
meter	100 cm or 10 dm
decimeter	10 cm
centimeter	10 mm
foot	12 in.
yard	3 ft or 36 in.
mile	5,280 ft or 1,760 yd
10 cm is about 4 in.	

Time	
year	365 or 366 days
year	about 52 weeks
year	12 months
month	28, 29, 30, or 31 days
week	7 days
day	24 hours
hour	60 minutes
minute	60 seconds

<	*is less than*
>	*is more than*
=	*is equal to*
=	*is the same as*

Money	
1¢, or $0.01	Ⓟ
5¢, or $0.05	Ⓝ
10¢, or $0.10	Ⓓ
25¢, or $0.25	Ⓠ
100¢, or $1.00	$1

Abbreviations	
kilometers	km
meters	m
centimeters	cm
miles	mi
feet	ft
yards	yd
inches	in.
tons	T
pounds	lb
ounces	oz
kilograms	kg
grams	g
decimeters	dm
millimeters	mm
pints	pt
quarts	qt
gallons	gal
liters	L
milliliters	mL

Capacity
1 pint = 2 cups
1 quart = 2 pints
1 gallon = 4 quarts
1 liter = 1,000 milliliters

Date ______________________ Time ______________________

LESSON 8·2

Fact Power Table

0 + 0	0 + 1	0 + 2	0 + 3	0 + 4	0 + 5	0 + 6	0 + 7	0 + 8	0 + 9
1 + 0	1 + 1	1 + 2	1 + 3	1 + 4	1 + 5	1 + 6	1 + 7	1 + 8	1 + 9
2 + 0	2 + 1	2 + 2	2 + 3	2 + 4	2 + 5	2 + 6	2 + 7	2 + 8	2 + 9
3 + 0	3 + 1	3 + 2	3 + 3	3 + 4	3 + 5	3 + 6	3 + 7	3 + 8	3 + 9
4 + 0	4 + 1	4 + 2	4 + 3	4 + 4	4 + 5	4 + 6	4 + 7	4 + 8	4 + 9
5 + 0	5 + 1	5 + 2	5 + 3	5 + 4	5 + 5	5 + 6	5 + 7	5 + 8	5 + 9
6 + 0	6 + 1	6 + 2	6 + 3	6 + 4	6 + 5	6 + 6	6 + 7	6 + 8	6 + 9
7 + 0	7 + 1	7 + 2	7 + 3	7 + 4	7 + 5	7 + 6	7 + 7	7 + 8	7 + 9
8 + 0	8 + 1	8 + 2	8 + 3	8 + 4	8 + 5	8 + 6	8 + 7	8 + 8	8 + 9
9 + 0	9 + 1	9 + 2	9 + 3	9 + 4	9 + 5	9 + 6	9 + 7	9 + 8	9 + 9

Date Time

U.S. Traditional Addition 1

Algorithm Project 1

Write a number model to show your ballpark estimate.
Use any strategy to solve the problem.

1. Jonah had a garage sale. He earned $52 on Saturday and $34 on Sunday. How much money did Jonah earn at the garage sale?

Ballpark estimate: ____________

$ ________

Write a number model to show your ballpark estimate.
Use U.S. traditional addition to solve each problem.

2. Ballpark estimate:

____

$$\begin{array}{r} 36 \\ +\ 23 \\ \hline \end{array}$$

3. Ballpark estimate:

____

$$\begin{array}{r} 19 \\ +\ 76 \\ \hline \end{array}$$

4. Ballpark estimate:

____

$$\begin{array}{r} 198 \\ +\ 46 \\ \hline \end{array}$$

5. Ballpark estimate:

____

$$\begin{array}{r} 187 \\ +\ 123 \\ \hline \end{array}$$

Date Time

U.S. Traditional Addition 2

Algorithm Project 1

Write a number model to show your ballpark estimate.
Use U.S. traditional addition to solve each problem.

1. Nia collects seashells. She had 29 seashells. She found 47 more seashells at the beach. How many seashells does Nia have now?

   Ballpark estimate: ______________

   ________ seashells

2. Ballpark estimate:

   ____

$$\begin{array}{r} 41 \\ +\ 38 \\ \hline \end{array}$$

3. Ballpark estimate:

   ____

$$\begin{array}{r} 96 \\ +\ 63 \\ \hline \end{array}$$

4. Ballpark estimate:

   ____

$$\begin{array}{r} 187 \\ +\ \ 78 \\ \hline \end{array}$$

5. Ballpark estimate:

   ____

$$\begin{array}{r} 208 \\ +\ 133 \\ \hline \end{array}$$

 Go to www.everydaymathonline.com for additional practice pages.

Date Time

U.S. Traditional Addition 3

Algorithm Project 1

Write a number model to show your ballpark estimate.
Use U.S. traditional addition to solve the problem.

1. There are two second grade classes at Park School. One class has 34 children. The other has 29 children. How many children are in second grade at Park School?

   Ballpark estimate: ______________________

   ________ children

2. Write a number story for 18 + 57.
   Solve your number story.

   ____

   ____

   ____

   ____

   ____

Fill in the missing digits in the addition problems.

3.

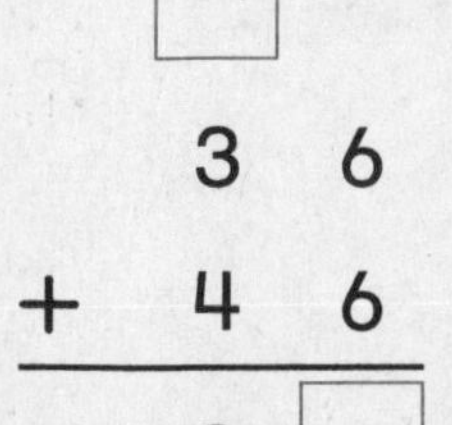

4.

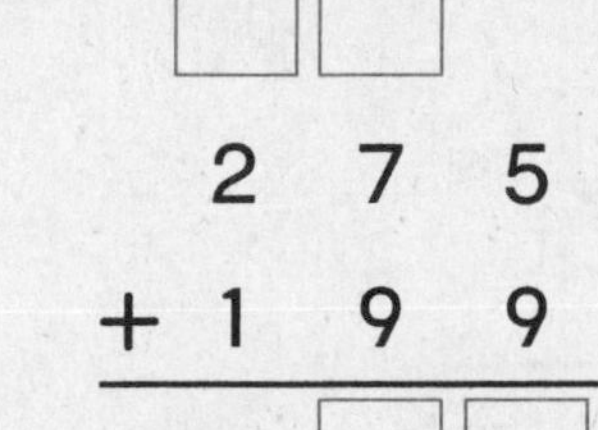

Date Time

U.S. Traditional Addition 4

Algorithm Project 1

Write a number model to show your ballpark estimate.
Use U.S. traditional addition to solve the problem.

1. Sam had $66 in his piggy bank. He earned $14 babysitting this weekend. How much money does Sam have now?

Ballpark estimate: ______________________

$________

2. Write a number story for 35 + 29. Solve your number story.

____

____

____

____

____

Fill in the missing digits in the addition problems.

3.

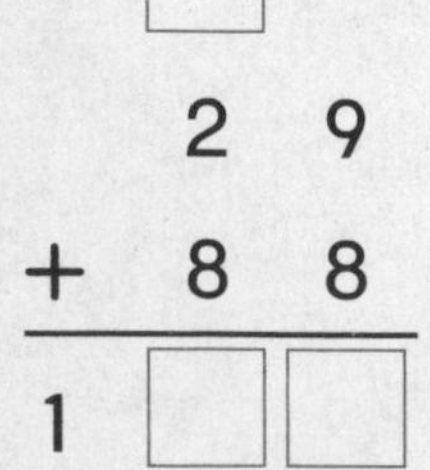

```
   □
   2 9
+  8 8
 1 □ □
```

4.

```
   □ □
   4 1 3
+    8 8
   5 □ 1
```

 Go to www.everydaymathonline.com for additional practice pages.

Date Time

U.S. Traditional Subtraction 1

Algorithm Project 2

Write a number model to show your ballpark estimate.
Use any strategy to solve the problem.

1. Jakob has 93 red blocks and 58 blue blocks in his toy box. How many more red blocks than blue blocks are in the toy box?

Ballpark estimate: ______________________

__________ red blocks

Write a number model to show your ballpark estimate.
Use U.S. traditional subtraction to solve each problem.

2. Ballpark estimate:

____

$$\begin{array}{r} 78 \\ -\ 37 \\ \hline \end{array}$$

3. Ballpark estimate:

____

$$\begin{array}{r} 96 \\ -\ 47 \\ \hline \end{array}$$

4. Ballpark estimate:

____

$$\begin{array}{r} 193 \\ -\ 76 \\ \hline \end{array}$$

5. Ballpark estimate:

____

$$\begin{array}{r} 252 \\ -\ 169 \\ \hline \end{array}$$

Date Time

U.S. Traditional Subtraction 2

Algorithm Project 2

Write a number model to show your ballpark estimate.
Use U.S. traditional subtraction to solve each problem.

1. Rosa had 56 baseball cards. She gave 17 cards to her friend Andre. How many cards does she have left?

Ballpark estimate: ____________________

________ baseball cards

2. Ballpark estimate:

____

$$\begin{array}{r} 89 \\ -\ 71 \\ \hline \end{array}$$

3. Ballpark estimate:

____

$$\begin{array}{r} 46 \\ -\ 28 \\ \hline \end{array}$$

4. Ballpark estimate:

____

$$\begin{array}{r} 154 \\ -\ 45 \\ \hline \end{array}$$

5. Ballpark estimate:

____

$$\begin{array}{r} 548 \\ -\ 259 \\ \hline \end{array}$$

Date Time

U.S. Traditional Subtraction 3

Algorithm Project 2

Write a number model to show your ballpark estimate.
Use U.S. traditional subtraction to solve the problem.

1. There are 84 children who ride a bus to school. 65 of them are boys. How many girls ride a bus to school?

   Ballpark estimate: ______________________

   ________ girls

2. Write a number story for 79 − 68. Solve your number story.

   ____

   ____

   ____

   ____

   ____

Fill in the missing numbers in the subtraction problems.

3.

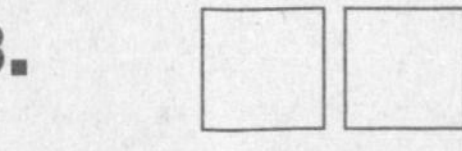

```
   ☐ ☐
   5̸ 1̸4
 −  2  5
 -------
    2  ☐
```

4.

```
     ☐ ☐
  1  9̸  1̸
 −   4  6
 ---------
  ☐  4  5
```

Date Time

U.S. Traditional Subtraction 4

Algorithm Project 2

Write a number model to show your ballpark estimate.
Use U.S. traditional subtraction to solve the problem.

1. Jenna's book has 46 pages. Shen's book has 97 pages. How many more pages are in Shen's book?

Ballpark estimate: ______________________

__________ pages

2. Write a number story for 37 − 18.
Solve your number story.

____

____

____

____

____

Fill in the missing numbers in the subtraction problems.

3.

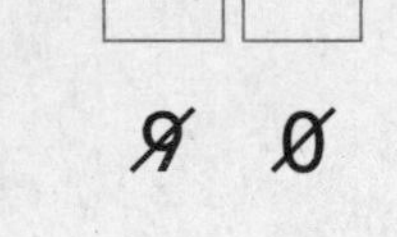

$$\begin{array}{r} \square\ \square \\ \not{9}\ \not{0} \\ -\ 6\ 4 \\ \hline 2\ \square \end{array}$$

4.

$$\begin{array}{r} \square\ \square\ \ \\ 3\ \not{7}\ \not{3} \\ -\ 1\ 3\ 6 \\ \hline \square\ \square\ 7 \end{array}$$

 Go to www.everydaymathonline.com for additional practice pages.

Date Time

LESSON 8•5

Fraction Cards

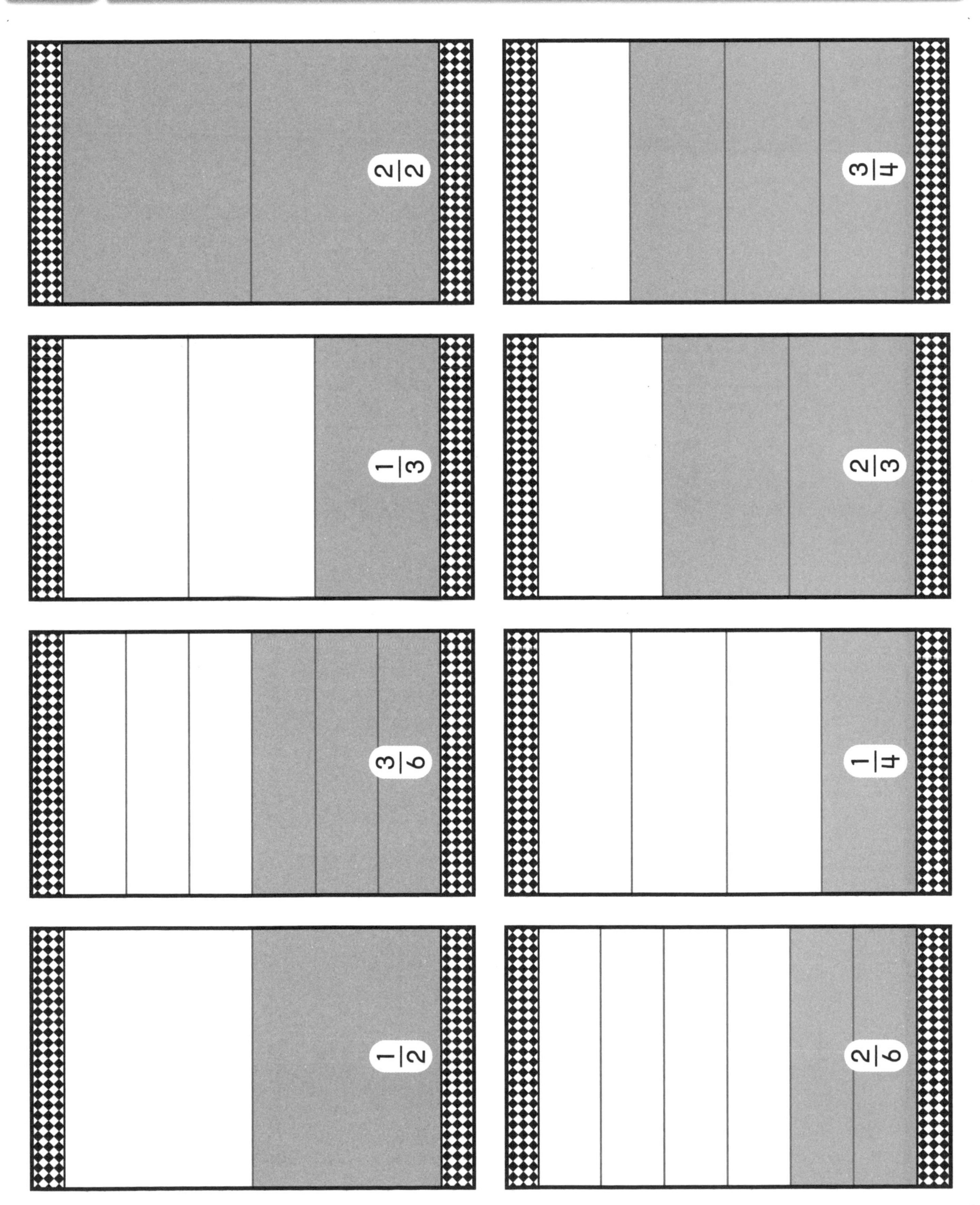

Date Time

LESSON 8•5

Fraction Cards

$\frac{3}{4}$ $\frac{2}{2}$

$\frac{2}{3}$ $\frac{1}{3}$

$\frac{1}{4}$ $\frac{3}{6}$

$\frac{2}{6}$ $\frac{1}{2}$

Date Time

LESSON 8•5

Fraction Cards

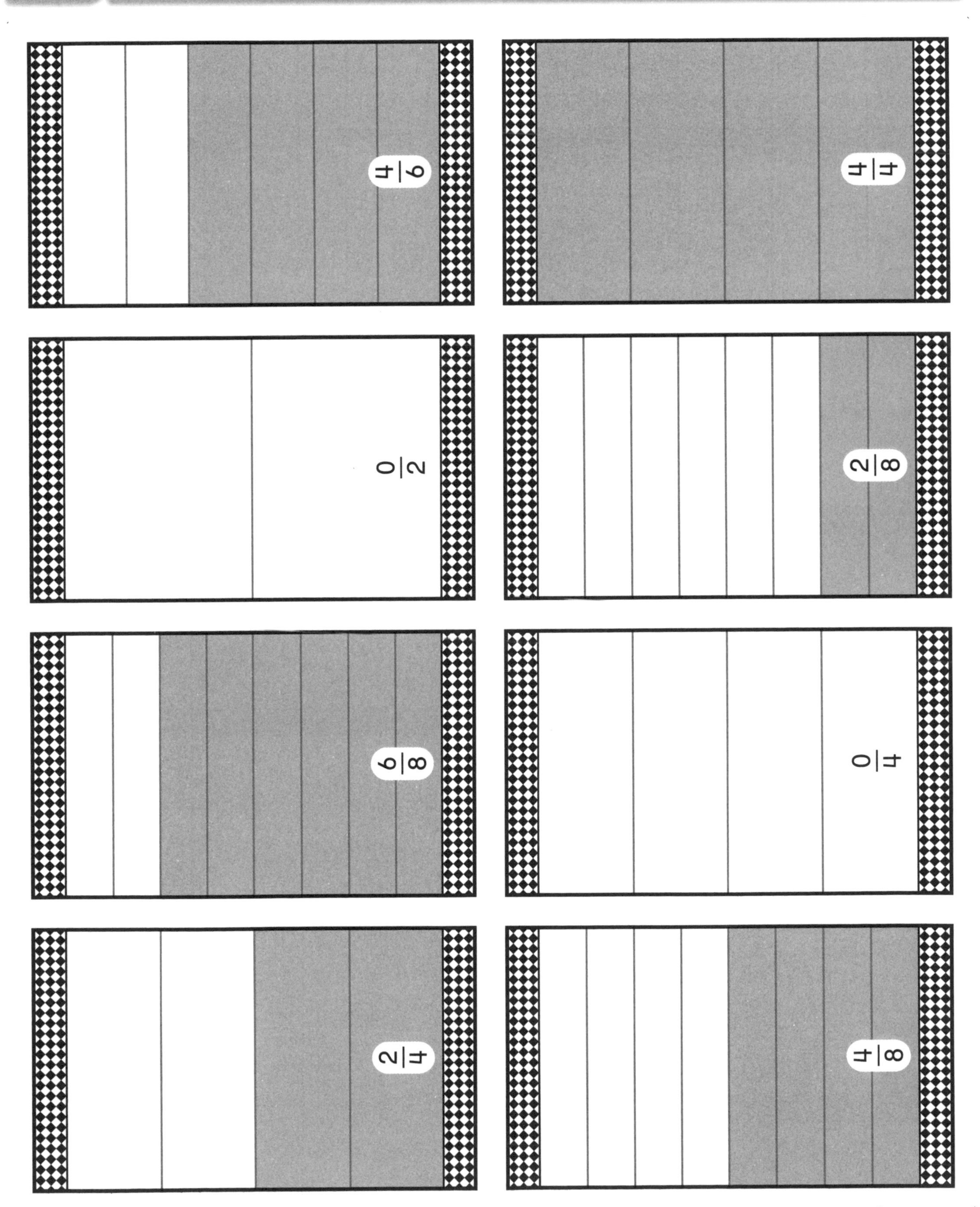

Date Time

Fraction Cards

$\frac{4}{4}$ $\frac{4}{6}$

$\frac{2}{8}$ $\frac{0}{2}$

$\frac{0}{4}$ $\frac{6}{8}$

$\frac{4}{8}$ $\frac{2}{4}$

Date Time

LESSON 11·7

×, ÷ Fact Triangles 1

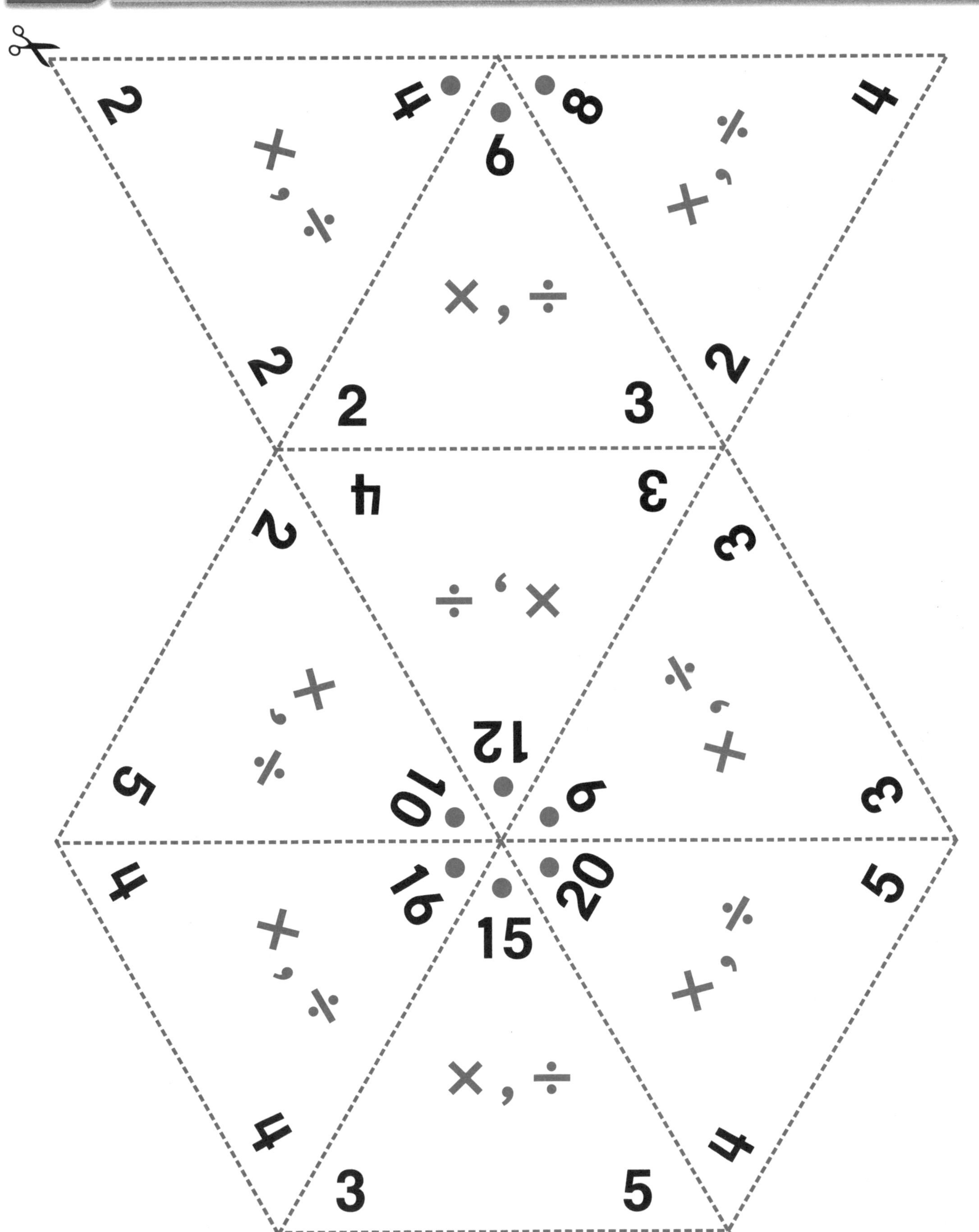

Date Time

LESSON 11•7

×, ÷ Fact Triangles 2

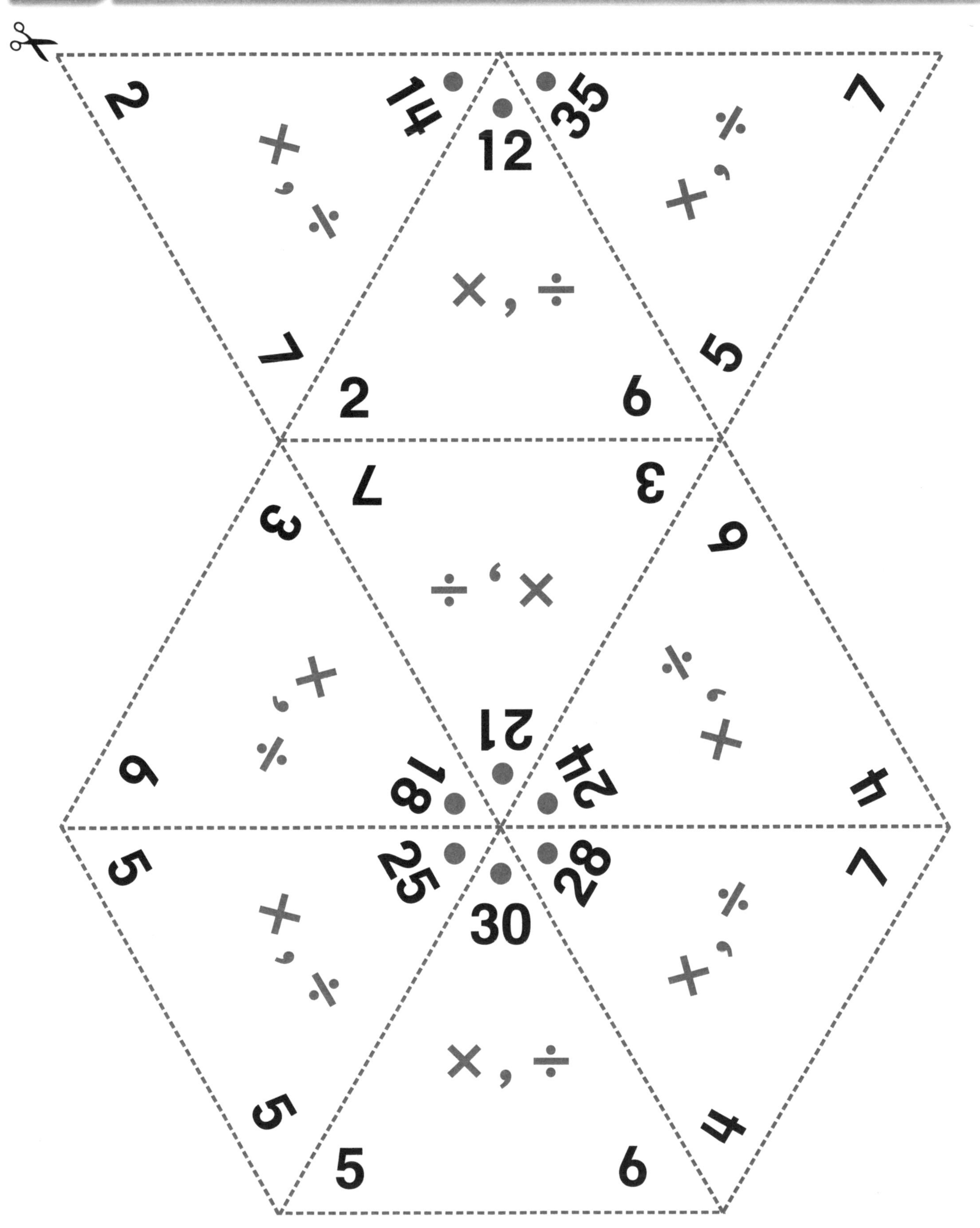

Date Time

×, ÷ Fact Triangles 3

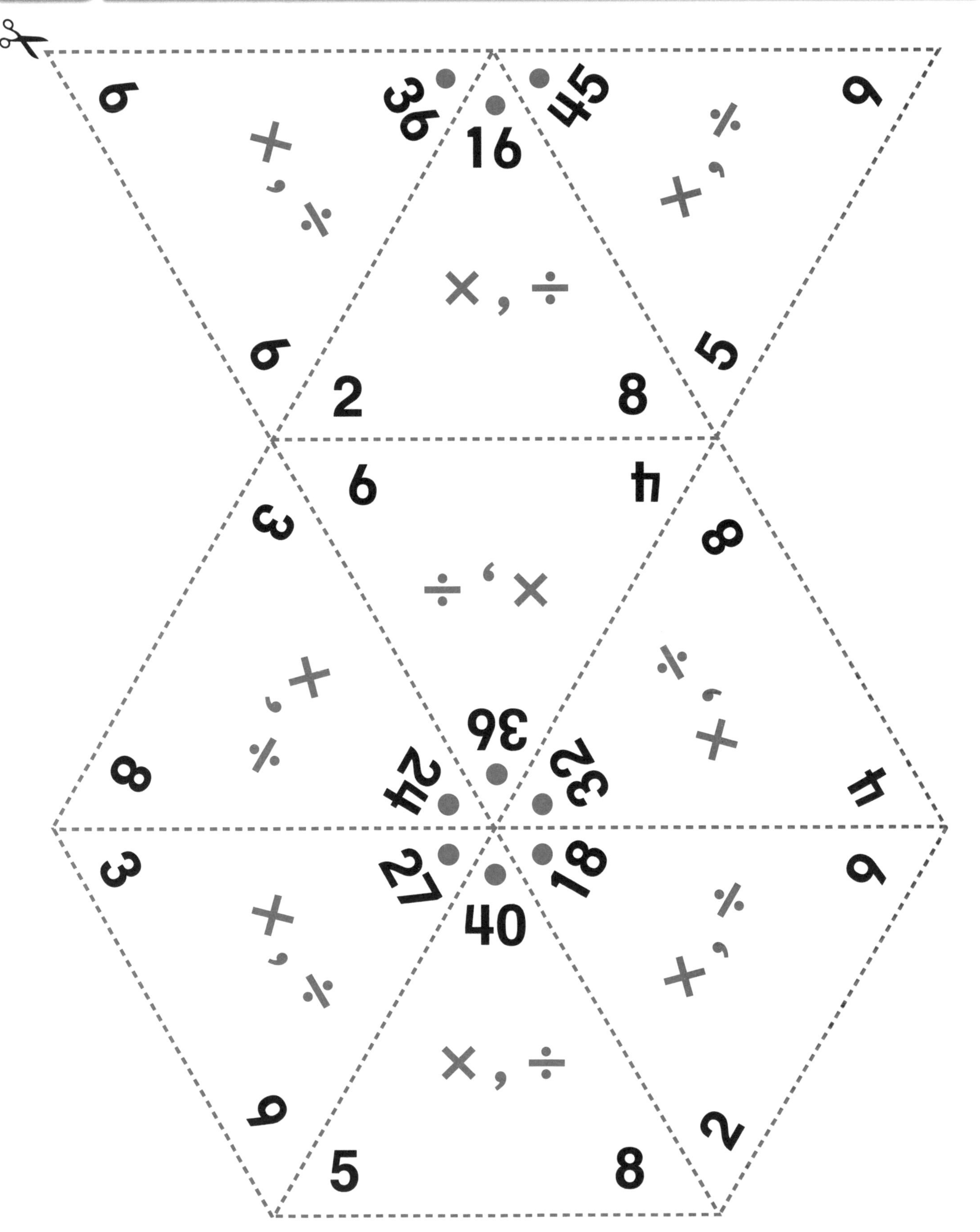

Date Time

LESSON 11·9

×, ÷ Fact Triangles 4

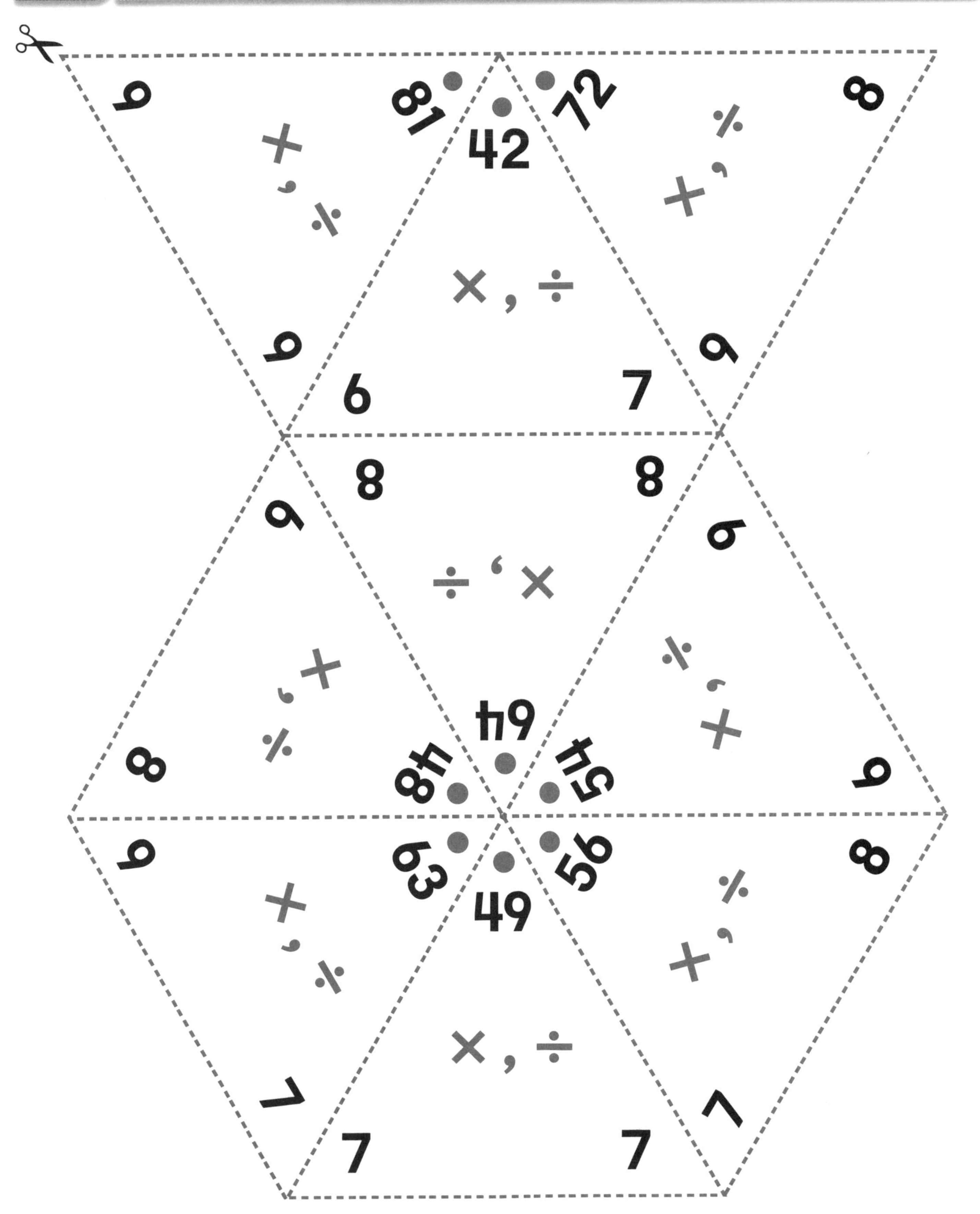